AF689843

Pearson Education

AP* Test Prep Series

AP* ENVIRONMENTAL SCIENCE

For:

ENVIRONMENTAL SCIENCE

TENTH EDITION

RICHARD T. WRIGHT

COURTNEY MAYER

Winston Churchill High School

* AP and Advanced Placement are registered trademarks of The College Board, which was not involved in the production of, and does not endorse, this product.

PEARSON Prentice Hall

Upper Saddle River, NJ 07458

Editor-in-Chief, Science: Nicole Folchetti
Publisher, Geosciences and Environment: Dan Kaveney
Project Manager: Crissy Dudonis
Senior Managing Editor, Science: Kathleen Schiaparelli
Assistant Managing Editor, Science: Gina M. Cheselka
Supplement Cover Manager: Paul Gourhan
Supplement Cover Designer: Victoria Colotta
Senior Operations Specialist: Alan Fischer
Director of Operations: Barbara Kittle

© 2008 Pearson Education, Inc.
Pearson Prentice Hall
Pearson Education, Inc.
Upper Saddle River, NJ 07458

All rights reserved. No part of this book may be reproduced, in any form or by any means, without permission in writing from the publisher.

Pearson Prentice Hall™ is a trademark of Pearson Education, Inc.

The author and publisher of this book have used their best efforts in preparing this book. These efforts include the development, research, and testing of the theories and programs to determine their effectiveness. The author and publisher make no warranty of any kind, expressed or implied, with regard to these programs or the documentation contained in this book. The author and publisher shall not be liable in any event for incidental or consequential damages in connection with, or arising out of, the furnishing, performance, or use of these programs.

Printed in the United States of America

10 9 8 7 6 5 4 3 2 1

ISBN 13: 978-0-13-196161-6

ISBN 10: 0-13-196161-6

Pearson Education Ltd., *London*
Pearson Education Australia Pty. Ltd., *Sydney*
Pearson Education Singapore, Pte. Ltd.
Pearson Education North Asia Ltd., *Hong Kong*
Pearson Education Canada, Inc., *Toronto*
Pearson Educación de Mexico, S.A. de C.V.
Pearson Education—Japan, *Tokyo*
Pearson Education Malaysia, Pte. Ltd.

TABLE OF CONTENTS

Courtney Mayer is a College Board Consultant in the Southwestern Region where she has conducted workshops throughout Texas and Oklahoma. She has taught in San Antonio, Texas for 15 years and has been teaching students AP Environmental Science for 7 years at Winston Churchill High School where enrollment in AP Environmental Science has grown to 235 students. She is a reader for the AP Environmental Science exam and earned her B.S. in Wildlife and Fisheries Science from Texas A&M University in 1992.

Introduction

Welcome to AP Environmental Science! This is an exciting course that will bring to life many topics that you will hear about on the news and in your daily lives. It is very hands on and applicable to the real world. It is my hope that you will leave this class with a better understanding and appreciation for the natural world around you while being able to make educated decisions about the way you live and conduct yourselves and your family. We live in a very fragile and beautiful world of which we need to be aware so as to be better stewards. I hope you enjoy this class and learn more than you ever expected!

For many of you, this is the first time you will have ever taken an AP course. Others of you may have a full load of AP courses and will be adding AP Environmental Science (APES) to your schedule. APES is perfect for either type of student. It is a great beginning AP course that is manageable for those who have never been exposed to the rigor of AP, and it is a wonderful learning experience to add for those with a full plate of AP courses. Many of my students come back to visit from college and have decided to major in some sort of Environmental Science: environmental law, environmental design, environmental engineering, etc. I hope you also will find a love and desire for this course and may want to continue to pursue further studies in this field.

Keys To Success

Get a calendar and get organized! The biggest mistake that new AP students make is missing important deadlines or losing important papers. My suggestion is to get a calendar and write down all assignment due dates, test dates, lab report write ups, etc. Then, get in the habit of checking that calendar many times a day. Get a notebook with dividers and get organized! Keep all your papers, notes, labs and tests. Many colleges want proof of the work you did in your high school APES course even if you make a 4 or 5 on the AP test. By keeping an organized notebook you can go to your college advisor and fight for credit if they are hesitant to grant it. If you have nothing you may earn nothing!

Study early and often! Another mistake students make is waiting until the last minute and then cramming for a test or for the APES exam. This almost never works. If you come home each day and go through the notes while they are fresh on your mind, you can remember them better and go in for tutoring help if there is something you don't understand. Teachers do not like it when a student shows up at their door the day before or the day of the test with a million questions. This just proves that the student is not prepared. Studies have shown that studying a little every day is more beneficial than last minute trying to cram it all in!

Read the book early! If you read before the day the teacher will be lecturing you will have a better understanding of the concepts before you walk into the classroom, you will be a more active learner, will be able to answer questions the teacher asks and will gain more out of the class. This textbook is incredibly written with easy to understand examples and helpful guest essays. Use it and benefit from it!

Look for the connections! Environmental Science is not a class that is to be taken in chunks. Soil is not independent of air quality issues but is directly connected to the problems we have with our water. Generally these topics are taught alone, but remember the connections. As you study one topic see how it relates to something you learned earlier in the year. I heard a wonderful example of this on a trip to Costa Rica. There is an island off the coast of Costa Rica that used to have some of the most beautiful coral reefs surrounding it. I asked our guide about scuba diving there and he told me not to bother, that the pigs have ruined the coral! Pigs??? How can pigs ruin a coral reef? Well, the natives brought pigs to the island as a source of food, and pigs dig in the soil and are very destructive of the ecosystem. As the pigs dug up the land, the land eroded and washed into the ocean. This sediment blocked out the sunlight and amount of photosynthesis that the coral were able to do. Now the coral reefs are dying because of the pigs. Look for these connections and your experience in this class and the way you look at your surroundings will change!

Be active in class and in the laboratory! Many labs in this class are done in groups. I see student after student sit back and let the other people in their group do all the work. How can you expect to understand the concepts your teacher is trying to teach if you are a passive learner? Get involved, get active and get working! I have a quote posted in my room that says, "Tell me and I will forget, show me and I may remember, involve me and I will understand." I cannot stress the importance of this in influencing the score you make on the AP exam. If you want a 5 you need to work for a 5 and sitting back and watching others work is not the way to get a 5 on this test!

Get connected with www.collegeboard.com! The College Board uploads every Free Response Question (FRQ) that has ever been asked on the AP test along with the rubric of how they graded it. I suggest starting about a month before the date of the APES exam to begin working through the released FRQ's. Test yourself by pretending it is the real exam, answering the question to the best of your ability and then grade yourself as if you were a grader for the test. I am a grader (reader) of the AP Environmental Science test and read so many tests of students that know the ideas but cannot get all the information written down in a form to earn them the credit they deserve. I find that if a question has two parts to it, students answer the first part and then forget to answer the second part, earning them half credit. By practicing the released exams and grading them yourself, you will learn to catch these simple mistakes which could make the difference between a passing or a failing APES exam!

Use the AP Environmental Science Test Prep Guide! I wrote the test prep guide to go with this textbook. As a teacher of AP Environmental Science, a consultant for the College Board and a grader for the AP Environmental Science exam, I have taken this textbook and pulled out the most important concepts that you need to review and study before the exam, along with multiple choice example questions and practice Free Response Questions. Use this tool to your advantage. It is an excellent book to remind you of concepts that you learned early in the year as well it has two full practice exams. The worst mistake and the biggest waste of money is to walk into the AP exam without reviewing the concepts you have learned this year. You will be wasting your time and money as well as the graders time. Review and study and you will be successful!

Good luck! I hope that you enjoy this class and do well on the exam. Earning college credit in high school allows you to cut tuition costs, allows you to lessen the class load you are taking in college which lowers you stress level and it also lets you take other classes that you would not have room for, or more importantly to graduate early if that is your goal. Maximizing the amount of AP credit you can earn is a win-win situation. I hope you learn to appreciate and protect the world around you and help to better the world for our children and grandchildren!! This world is fragile and precious and it takes everyone to help protect and preserve it!!!

Introduction: Toward a Sustainable Future

Easter Island teaches us many lessons:

- Humans can impact an entire environment, leading to unintended consequences.

- Populations can increase beyond the carrying capacity of the land. When this happens, the land can no longer provide adequate food for all the people.

1.1 The Global Environmental Picture

- The world's human population is over 6.6 billion in 2007 and has grown by 2 billion in the last 25 years.

- The Earth's human population is growing by over 76 million people each year.

- The U.N. Population Division estimates that world human population could be 9.1 billion by 2050.

- Ecosystems are declining around the world. Groundwater supplies are being depleted, agricultural soils degraded, oceans overfished, and forests cut faster than they can regrow.

- According to the Millennium Ecosystem Assessment, 60% of the ecosystems assessed are being degraded or used unsustainably.

- In 1987, concern about depletion of the stratospheric ozone layer led to the Montreal Protocol, an international action aimed at curbing pollution from the release of chlorofluorocarbon refrigerants into the atmosphere.

- In Kyoto, Japan in 1997, 166 nations met to negotiate a treaty to reduce emissions of carbon dioxide and other greenhouse gases to below 1990 levels by the year 2010. This treaty is known as the Kyoto Protocol and was set forth to address global climate change. (On February 14, 2007 169 nations had ratified this treaty.)

- Biodiversity, the total diversity of living things, is quickly being lost around the planet by, among other things, the growing human population and pollution.

1.2 Three Strategic Themes: Sustainability, Stewardship, and Science

Sustainability	A process that can be continued indefinitely without depletion of its required material or energy resources
Stewardship	An attitude of active care and concern for natural lands
Science and the Scientific Method	Gaining objective knowledge by a method that utilizes observations, hypotheses, testing, and theory

1.3 Three Integrative Themes: Ecosystem Capital, Policy/Politics, and Globalization

Ecosystem Capital	The sum of goods and services provided by natural and managed ecosystems, free of charge, and essential to human life and well-being
Policy/Politics	The human decisions that determine what happens to the natural world
Globalization	The interconnectedness of human activities, ideas, and cultures

The goal of the above six themes is a sustainable future.

1.4 The Environment in the 21st Century

- Many people are playing an important role in changing society's treatment of the Earth.
- Population growth rates are falling in many developing countries.
- Laws and treaties have reduced pollution significantly.
- Environmental degradation can be slowed down or reduced.

Ecosystems: What They Are

2.1 Ecosystems: A Description

Species	A population that can interbreed and produce fertile offspring
Populations	Groups of individuals of the same species that can interbreed with one another and that all live in one area
Community	Populations of all species that live and interact a given area at a particular time.
Ecosystem	Many communities of different species interacting together and the abiotic or nonliving environment (soil, water, air, etc.)
Ecotone	A transitional area between two adjacent ecosystems that has characteristics of both
Biome	A particular type of ecosystem (tropical rain forest, desert, tundra, etc.)

2.2 The Structure of Ecosystems

Primary Producer (autotroph)	Organisms that make their own food by photosynthesis. (plants, phytoplankton, algae and some bacteria)
Consumer (heterotroph)	Organisms that eat living or dead material
Predator	Organisms that kill and eat other animals
Prey	Organisms that are eaten by a predator
Detritivore	Organisms that feed on dead or dying organisms
Decomposer	A type of detritivore that feeds on leaves or rotting wood breaking them down into simpler inorganic molecules which are later recycled in the ecosystem (ex. fungi and bacteria)
Food chain	A diagram showing a series of organisms and what they eat (seed→mouse→hawk) A direct feeding relationship
Food web	Many interconnected food chains (transfer of energy from one species to another in the ecosystem)
Niche	A role an organism plays in its environment
Habitat	A place where an animal lives

- **Consumers** are divided into three different groups: *Primary consumers*—herbivores (plant eaters) (e.g., mice)

- *Secondary consumers*—animals that eat primary consumers (e.g., snakes)

- *Tertiary consumers*—animals that eat secondary consumers (e.g., hawks)

Figure 2.1

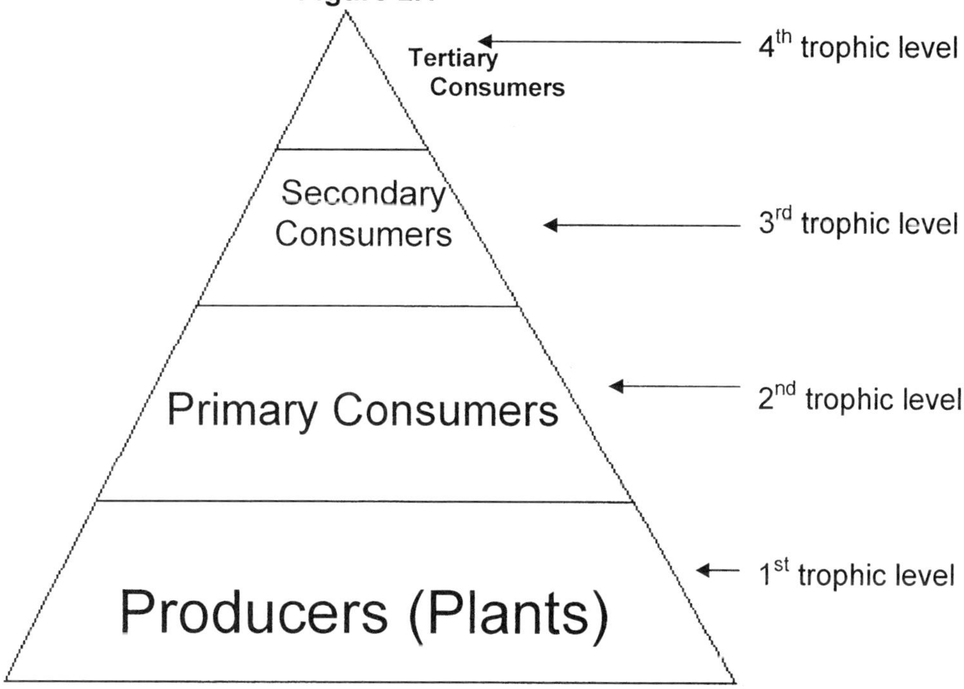

- As Figure 2.1 illustrates, each organism in a food chain can be placed in a **trophic pyramid**, with producers on the bottom and tertiary consumers on the top. Most trophic pyramids have three levels of consumers. As you move up the pyramid, you apply the **10% rule**, which states that 90% of all usable energy is lost to the environment as heat and only 10% of useable energy moves up through the trophic pyramid.

- The 10% rule applies because much of the food is converted to heat that the organism uses and because much of the biomass is consumed by the decomposers and never goes to the next level up.

- **Symbiosis** is two organisms living together. Not all symbiotic relationships are parasitic. Some organisms exhibit mutualism, which is when both organisms benefit from the symbiotic relationship.

2.3 From Ecosystems to Global Biomes

- **Climate** is determined by two factors, *temperature* and *precipitation*.

- Certain species can exist in different climates, depending on their **tolerance levels**. A species will be unable to live in an area that is outside of its tolerance level.

The world is divided into approximately six terrestrial biomes based on climate and tolerance levels. The chart below lists and summarizes these biomes along with the climate, precipitation levels, and soil types associated with each.

Biome	Characteristics
Desert	• Very dry; hot days and cold nights • Rainfall less than 10 in./yr • Soils thin and porous
Grasslands	• Seasonal rainfall, 10 to 60 in./yr • Fires frequent • Soils rich and often deep
Tropical Rain Forests	• Nonseasonal; annual average temperature 28°C • Rainfall frequent and heavy, average over 95 in./yr • Soils thin and poor in nutrients
Temperate Forests	• Seasonal; temperature below freezing in winter; summers warm and humid • Rainfall from 30–80 in./yr • Soils well developed
Coniferous Forests	• Seasonal; winters long and cold • Precipitation light in winter, heavier in summer • Soils acidic, much humus and litter
Tundra	• Bitter cold, except for an 8- to 10-week growing season with long days and moderate temperatures • Precipitation low • Soils thin and underlain with permafrost

Figure 2-2

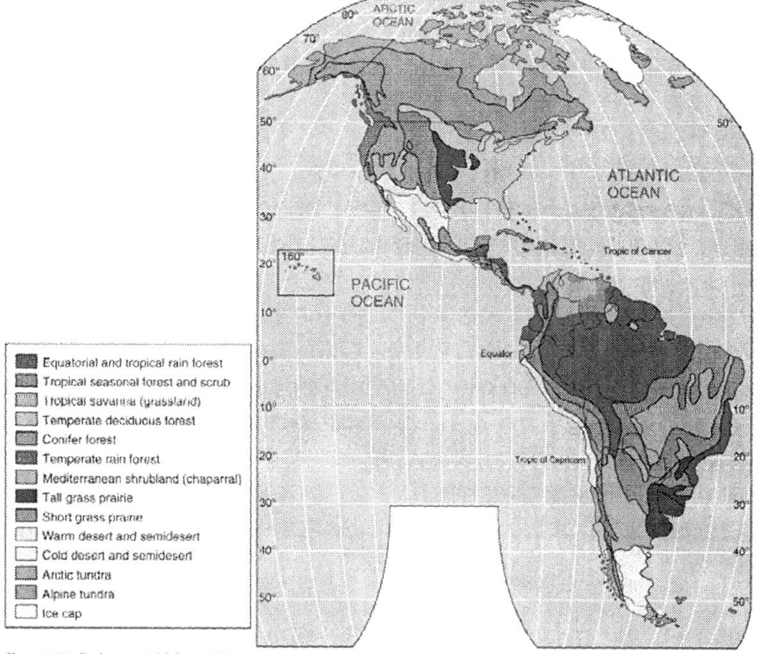

Figure 2-18a Environmental Science, 10/e
© 2008 Pearson Prentice Hall, Inc.

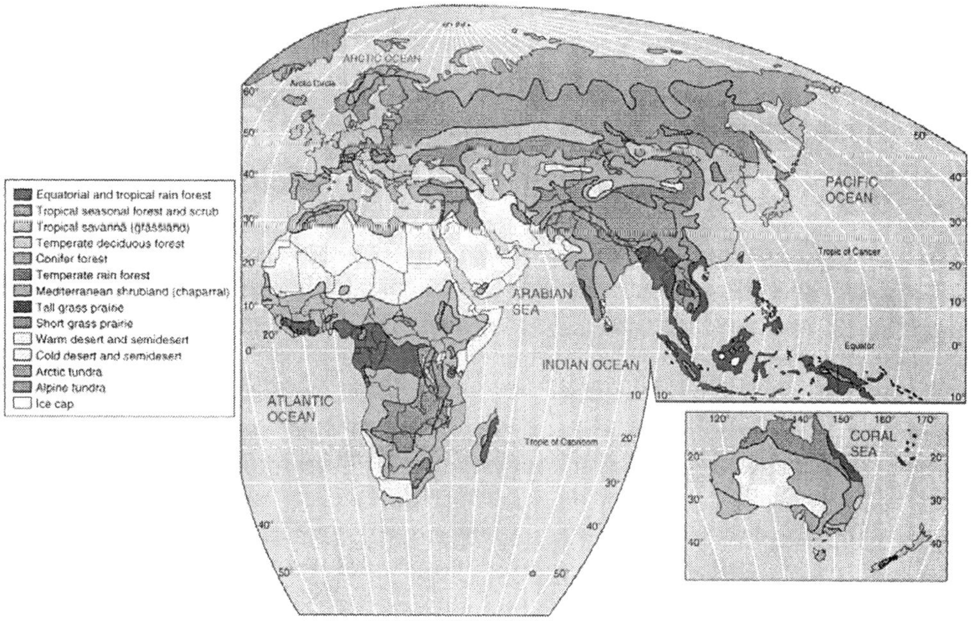

Figure 2-18b Environmental Science, 10/e
© 2008 Pearson Prentice Hall, Inc.

2.4 The Human Presence

Humans have a large impact on natural ecosystems. Learning to manage and restore these ecosystems is an important lesson in any environmental science course. Three past revolutions illuminate this relationship between humans and their environments.

- Neolithic Revolution (40,000 to 10,000 years ago): Humans first survived as hunter-gatherers, moving around as food shortages occurred. Then, during the Neolithic Revolution, humans developed agriculture in the Fertile Crescent 10,000 years ago, which provided a more reliable food supply and allowed for permanent settlements.

- Industrial Revolution (10,000 years ago to the 20th century): This revolution brought about global commerce, factories, large cities, and the use of fossil fuels. With the latter emerged large-scale pollution along with the use and extraction of raw materials from all over the world.

- Environmental Revolution (21st century): This revolution is based on the idea that a change is needed to take the world from its present state to one that is sustainable.

Multiple Choice Questions

1. Two of the most important factors that determine a climate are

    A. latitude and precipitation
    B. temperature and precipitation
    C. temperature and light
    D. the animals and plants that live there
    E. latitude and altitude

2. Tropical rain forests

    A. have deep, nutrient-rich soils
    B. have deep, nutrient-poor soils
    C. have shallow, nutrient-rich soils
    D. have shallow, nutrient-poor soils
    E. have a widely distributed soil pattern

3. A cattle egret standing on the back of a cow is an example of

    A. mutualism
    B. parasitism
    C. a predator-prey relationship
    D. commensalisms
    E. eutrophication

4. It is estimated that only ___________ % of available energy is passed from one level of the trophic pyramid to another.

    A. 1%
    B. 5%
    C. 10%
    D. 90%
    E. Answer cannot be determined based on information given.

5. Which of the following trophic levels has the most biomass?

    A. producer
    B. primary consumer
    C. secondary consumer
    D. decomposer
    E. tertiary consumer

6. The best example of a tertiary consumer would be a/an

    A. mouse
    B. grasshopper
    C. sheep
    D. cactus
    E. coyote

7. A decomposer would recycle nutrients back to a

    A. producer
    B. decomposer
    C. secondary consumer
    D. tertiary consumer
    E. primary consumer

Use the following diagram to answer questions 8–12:

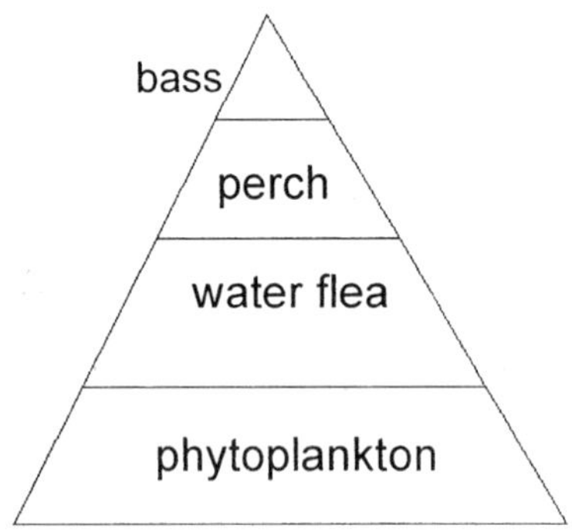

8. The largest amount of energy available in the pyramid is the

    A. phytoplankton
    B. bass
    C. perch
    D. water flea
    E. both A and C

9. What food chain is represented in the diagram?

    A. water flea → perch → bass→ phytoplankton
    B. perch → bass → phytoplankton
    C. phytoplankton → bass → perch → water flea
    D. phytoplankton →water flea → bass → perch
    E. phytoplankton → water flea → perch → bass

10. According to the 10% rule, to support 100 pounds of bass requires

A. 100 pounds of perch
B. 90 pounds of perch
C. 1000 pounds of water fleas
D. 10,000 pounds of water fleas
E. 900,000 pounds of phytoplankton

11. Which biome would be most affected by a large decrease in precipitation?

A. desert
B. tundra
C. coniferous forest
D. temperate forest
E. grassland

12. All of the following are examples of heterotrophs *except*

A. vulture
B. bacteria
C. moss
D. human
E. mushroom

13. The Sahel region of Africa, where the Sahara Desert meets the grassland, best describes a

A. population
B. association
C. biosphere
D. ecotone
E. ecosystem

14. All of the following are examples of abiotic factors *except*

A. temperature
B. pH
C. wind
D. salinity
E. vegetation

15. Which of the following would help to restore biodiversity?

A. increasing living standards in the developing world
B. deregulating pollution controls
C. introducing nonnative species into an environment
D. maintaining current levels of production and consumption
E. protecting ecosystems

16. If you travel south from Canada to Costa Rica, you would pass through many biomes. This change in biomes is mainly the result of

A. altitude
B. human factors
C. increase in temperature and precipitation
D. ocean currents
E. succession

Free Response Questions

Use the biomass pyramid below to answer the following questions.

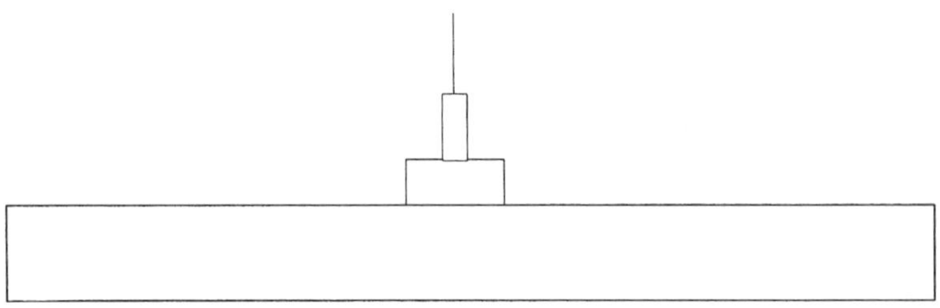

1. Explain why the pyramid is shaped as it is.

2. Assume the biomass of the organism at the bottom of the tropic pyramid is 5,000 pounds. **Diagram** the food pyramid, and **label** the approximate mass of each of the remaining levels.

3. Define and give an example of an herbivore, carnivore, and omnivore.

4. What is biomass, and how is it measured?

Multiple Choice Answers and Explanations

1. B. Climate is determined by two factors, *temperature* and *precipitation.*

2. D. Tropical rain forests have an annual average temperature of 28ºC; a frequent and heavy rainfall that averages over 95 inches per year; and thin and nutrient-poor soils.

3. A. Symbiosis is two organisms living together. Mutualism is when the two organisms both benefit from the symbiotic relationship, like in the case of the cattle egret and the cattle.

4. C. As you move up the trophic pyramid, you apply the 10% rule, which states that 90% of all usable energy is lost to the organism below and only 10% moves up through the trophic pyramid.

5. A. The biomass of all the organisms at each trophic level can be estimated by collecting and weighing suitable samples. In terrestrial ecosystems, the biomass is roughly 90% less at each higher trophic level. Thus, the bottom of the trophic pyramid (the producers) has the most biomass.

6. E. A tertiary consumer, such as a coyote, eats secondary consumers. Primary consumers –are herbivores or plant eaters (e.g., mouse), and secondary consumers are animals that eat primary consumers (e.g. snake).

7. A. As organic matter is broken down (decomposed), its chemical elements are released back to the environment, where, in the inorganic state, they may be reabsorbed by autotrophs (producers).

8. A. The biomass of all the organisms at each trophic level can be estimated by collecting and weighing suitable samples. In terrestrial ecosystems, the biomass is roughly 90% less at each higher trophic level. Therefore, phytoplankton is similar in an aquatic system to plants in a terrestrial ecosystem.

9. E. A food chain is the transfer of energy and material through a series of organisms as each one is fed upon by the organism higher up in the trophic pyramid. To answer this question, you would look at the diagram and, starting at the bottom, draw arrows from the producer to the consumer.

10. D. If there is 100 pounds of biomass in the bass population, there would be 1,000 pounds of biomass for the perch population, 10,000 pounds of biomass for the water flea population, and 100,000 pounds of biomass for the phytoplankton population.

11. D. Temperate forests require 30–80 inches of rain per year to support large trees, the greatest precipitation level of all the biomes listed. Tropical rain forests (not listed as a choice) require an even greater amount of rainfall than temperate forests.

12. C. Heterotrophs eat living or dead material. All choices are living or dead material except limestone.

13. D. An ecotone, such as the Sahel region of Africa, is a transitional area between two ecosystems that has characteristics of both.

14. E. Abiotic means nonliving. Vegetation is a biotic (living) factor.

15. E. All the choices would decrease biodiversity except for protecting ecosystems.

16. C. As you move from the poles to the equator, the amount of rainfall generally increases, and the temperature tends to increase. As you go from Canada to Costa Rica, the change in biomes is mainly the result of an increase in temperature and precipitation.

Free Response Answers

1. The pyramid shows the decreasing available amount of biomass at each level of the trophic pyramid.

2. Tertiary Consumer 5 lbs

Secondary Consumer 50 lbs

Primary Consumer 500 lbs

Producer
5,000 lbs

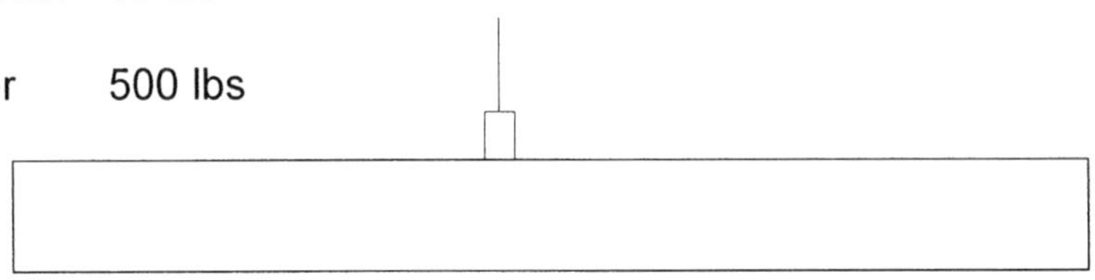

3. An herbivore is an organism that eats plants. An example of an herbivore is a cow. A carnivore is an animal that eats other animals. An example of a carnivore is a wolf. An omnivore is an organism that eats both plants and animals. A human is an example of an omnivore.

4. Biomass is the mass of biological material. It usually refers to the total mass of a particular group or category, for example, the biomass of producers. Biomass is measured by drying and weighing the biological material.

Ecosystems: How They Work

3.1 Matter, Energy, and Life

Biosphere	The part of the Earth that is occupied by organisms
Atmosphere (Air)	A mixture of three gases (nitrogen, oxygen, carbon dioxide) and a few trace gases
Hydrosphere (Water)	A major source of hydrogen bonded with oxygen to form water
Lithosphere (Land)	All other elements used by living organisms and the 72 elements not needed in the form of rock and soil

Organic molecules	Carbon-based molecules that make up the tissues of living organisms
Inorganic molecules	Molecules with neither carbon-carbon nor carbon-hydrogen bonds
Matter	Anything that occupies space and has mass
Encrgy	The ability to move matter (light, heat, movement, electricity)
Kinetic energy	Energy in action or in motion
Potential energy	Energy in storage
First law of thermodynamics (law of conservation of energy)	Energy is neither created nor destroyed, but it may be converted from one form to another.
Second law of thermodynamics	In any energy conversion, some usable energy is lost mostly as heat (entropy).

3.2 Energy Flow in Ecosystems

As energy flows from one trophic level to the next, only a small fraction is passed on. This reduction in energy is because

- much of the first trophic level is biomass and is not consumed.
- much of what is consumed is lost to the environment as heat.
- some of what is consumed is undigested and passed through the organism.

Thus, 90% of the available energy remains at the lower trophic level as it moves from one trophic level to the next. This 10% rule is why carnivores are less abundant than herbivores and why carnivores that eat other carnivores are the least abundant.

3.3 The Cycling of Matter in Ecosystems

The carbon, phosphorus, and nitrogen cycles illustrate the law of conservation of matter.

The Carbon Cycle: Carbon is found in all three "spheres:" the air, the land, and the water. As shown in Figure 3–16, carbon is taken out of the air by plants during photosynthesis and is returned to the air by cellular respiration. Carbon is also cycled through the tissues (biomass) of all organisms in the ecosystem.

Humans are diverting or canceling 40% of terrestrial production by burning fossil fuels and by destroying forests. Humans have increased atmospheric carbon dioxide by 35% over pre-industrial levels.

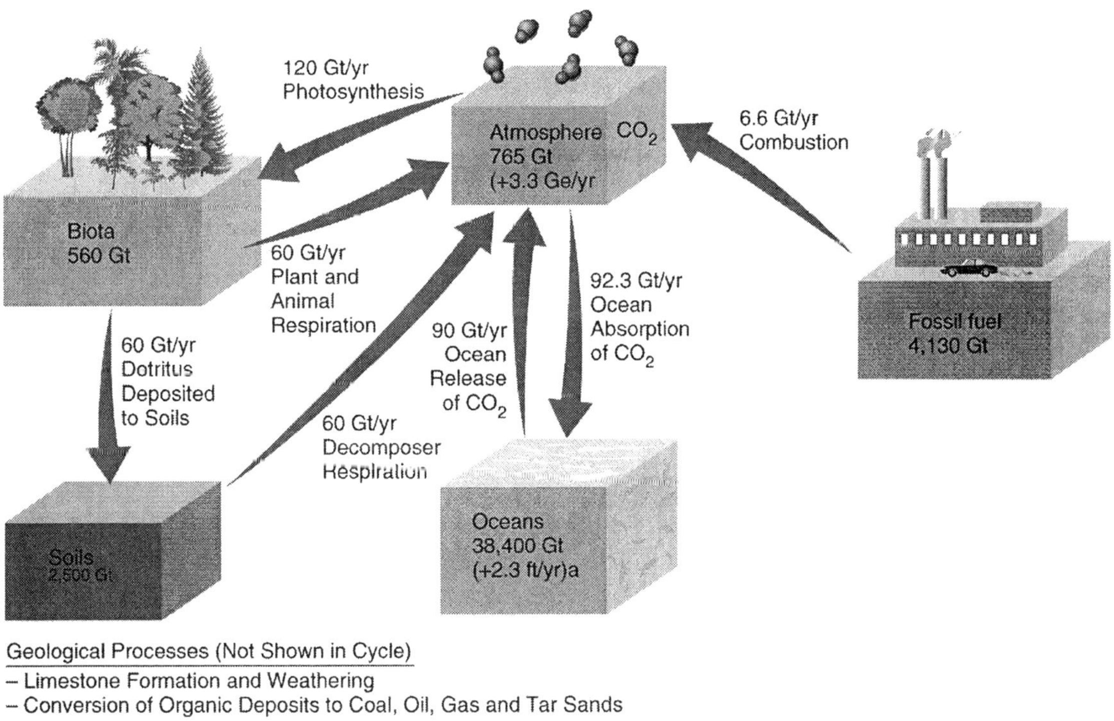

Figure 3-16 Environmental Science, 10/e
© 2008 Pearson Prentice Hall, Inc.

The Phosphorus Cycle: Unlike carbon, the phosphorus cycle does not have a gaseous phase, so phosphorus is recycled only if the wastes containing it are deposited in the ecosystem from which it came (Figure 3–17), thus acting as a limiting factor to plants.

Humans have impacted the phosphorus cycle by using phosphorus-containing fertilizers. These fertilizers runoff the land into the waterways. A severe water pollution problem, known as *eutrophication,* can result from dissolved phosphates.

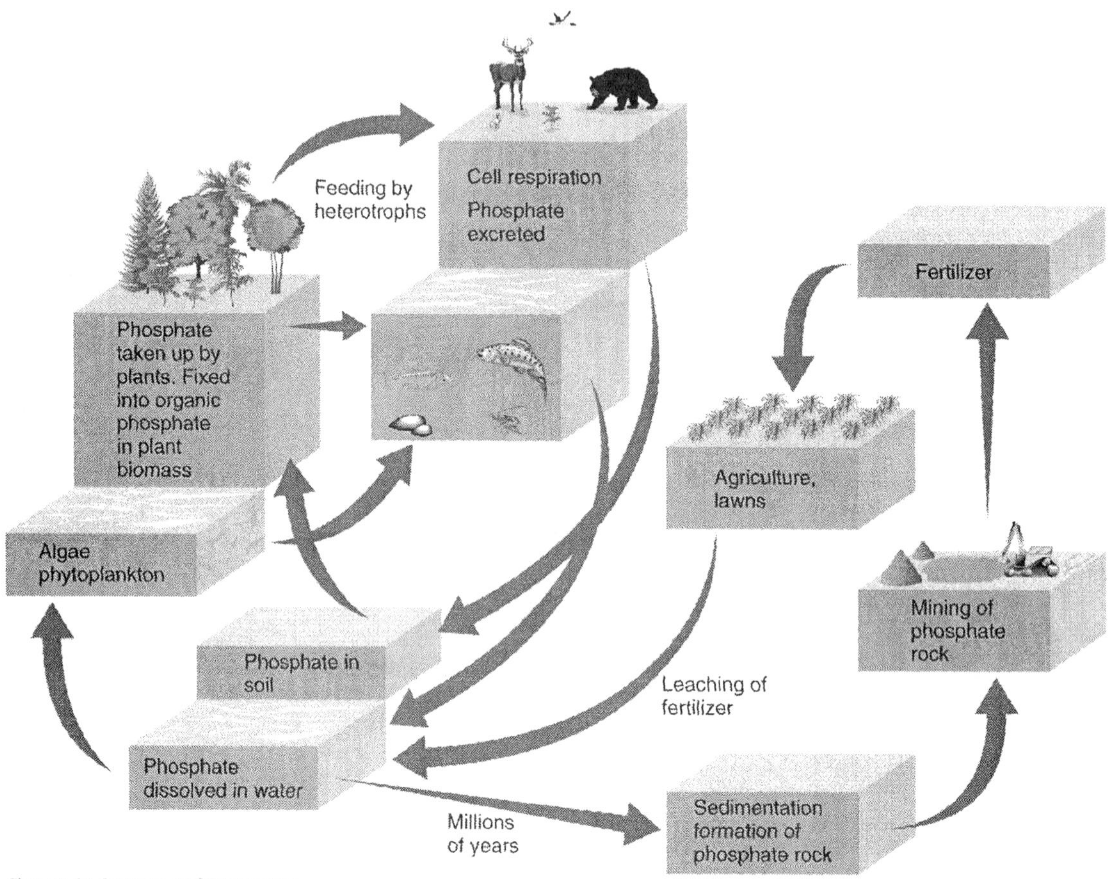

Figure 3-17 Environmental Science, 10/e
© 2008 Pearson Prentice Hall, Inc.

The Nitrogen Cycle: The main reservoir of nitrogen is the air, which is about 78% nitrogen gas. Plants cannot use nitrogen from the air; they must get it out of the soil. A number of bacteria and cyanobacteria can convert nitrogen gas to ammonium through a process called *nitrogen fixation.* Some plants like legumes have symbiotic bacteria living on their roots that can "fix" nitrogen and return it back to the food web. As illustrated in Figure 3–18, **denitrification** occurs when certain microbes use nitrate as a substitute for oxygen, thus returning nitrogen back to the atmosphere.

Humans impact the nitrogen cycle by planting many legume crops. Too many legume crops accelerates the normal rate of nitrogen fixation. The burning of fossil fuels also adds nitrogen to the ecosystem, and this added nitrogen can runoff into surface waters, causing algal blooms in waterways.

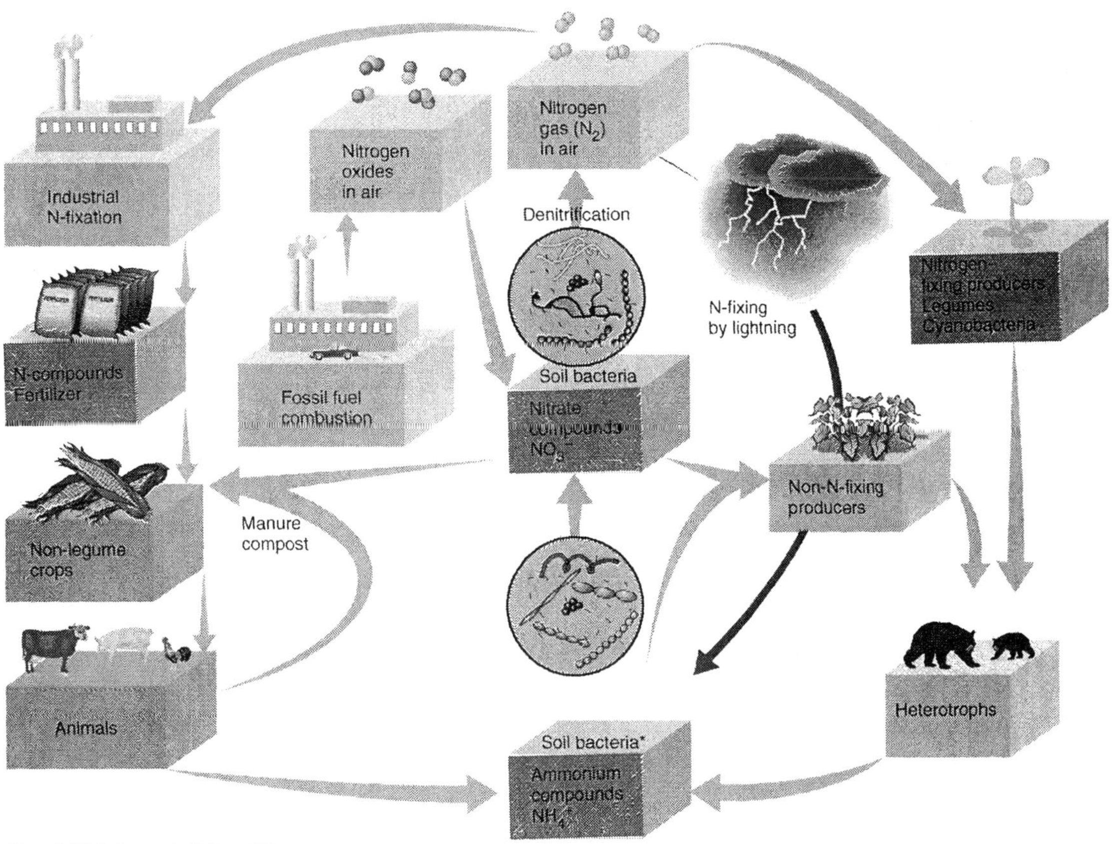

Figure 3–18 Environmental Science, 10/e
© 2008 Pearson Prentice Hall, Inc.

Multiple Choice Questions

1. Which major cycle does not contain a gaseous phase?

    A. nitrogen cycle
    B. phosphorous cycle
    C. carbon cycle
    D. water cycle
    E. none of the above

2. The most abundant element in the Earth's atmosphere is

    A. neon
    B. carbon dioxide
    C. oxygen
    D. nitrogen
    E. water vapor

3. The process by which microbes recycle nitrogen gas back into the atmosphere by using nitrate as a substitute for oxygen is called

    A. denitrification
    B. nitrogen fixation
    C. industrial fixation
    D. nitrification
    E. ammonification

For questions 4–8, refer to the following concepts related to energy.

A. first law of thermodynamics
B. second law of thermodynamics
C. entropy
D. kinetic energy
E. potential energy

4. Which of the concepts refers to energy in storage?

5. Which of the concepts states that energy can be converted from one form to another?

6. Which of the concepts refers to the loss of usable energy to heat?

7. Which of the concepts refers to energy in action or in motion?

8. Which of the concepts states that some usable energy is always lost in any conversion?

9. The part of the earth that is occupied by living things is the

    A. lithosphere
    B. atmosphere
    C. hydrosphere
    D. transophere
    E. biosphere

10. Matter is

    A. anything with mass and takes up spacc
    B. anything with the ability to do work
    C. made up of atoms
    D. two or more different kinds of atoms bonded together
    E. anything with the ability to be changed in form or shape

11. All forms of water make up the

    A. lithosphere
    B. atmosphere
    C. hydrosphere
    D. transophere
    E. biosphere

12. Global warming has been linked to humans altering which cycle?

    A. nitrogen cycle
    B. phosphorous cycle
    C. carbon cycle
    D. water cycle
    E. none of the above

13. Carbon dioxide is being added to the Earth's atmosphere in all of the following ways *except* for

    A. combustion of fossil fuels
    B. cellular respiration
    C. volcanism
    D. decomposition
    E. photosynthesis

14. Nodules would most likely be found on the roots of what plant species?

A. grasses
B. oaks
C. legumes
D. daisies
E. chrysanthemums

Free Response Questions

1. Using a clearly labeled diagram, explain the carbon cycle, the phosphorus cycle, or the nitrogen cycle. Make sure you show how the element is transported in the ecosystem.

2. Explain how the carbon cycle illustrates the law of conservation of matter.

Multiple Choice Answers and Explanations

1. B. Unlike the carbon cycle or the nitrogen cycle, the phosphorus cycle does not have a gaseous phase. Phosphorus is recycled only if the wastes containing it are deposited in the ecosystem from which it came.

2. D. Seventy-eight percent of the Earth's atmosphere is nitrogen.

3. A. Denitrification is a process in which certain microbes use nitrate as a substitute for oxygen, thus recycling nitrogen gas back to the atmosphere.

4. E. Potential energy is energy in storage.

5. A. The first law of thermodynamics states that energy is neither created nor destroyed, but energy may be converted from one form to another.

6. C. Entropy is the loss of usable heat to energy.

7. D. Kinetic energy is energy in action or in motion.

8. B. The second law of thermodynamics states that in any energy conversion, some usable energy is lost.

9. E. The biosphere is the part of the earth that is occupied by living things.

10. A. Matter is anything with mass and takes up space.

11. C. The hydrosphere is the part of the earth made of water.

12. C. Global warming results from an excess of carbon dioxide in the atmosphere. This excess of carbon dioxide is linked to humans burning of fossil fuels.

13. E. Photosynthesis does not add carbon to the atmosphere.

14. C. Nodules would be found on the roots of legumes. Bacteria living in nodules on the roots of legumes are able to fix nitrogen.

Free Response Answers

1. Review the diagram of the element cycle that you chose. Make sure you listed correctly each step in the cycle.

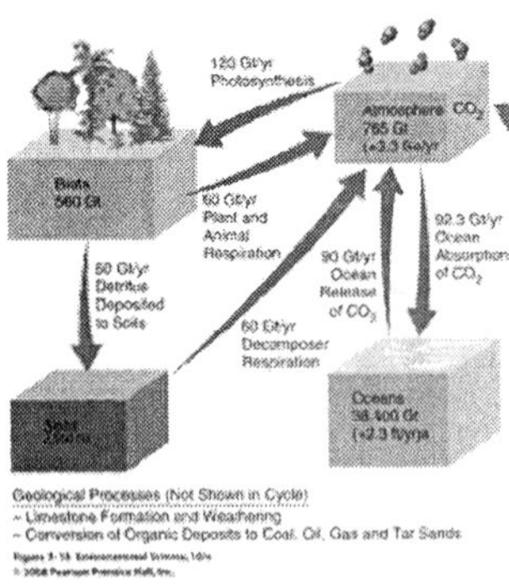

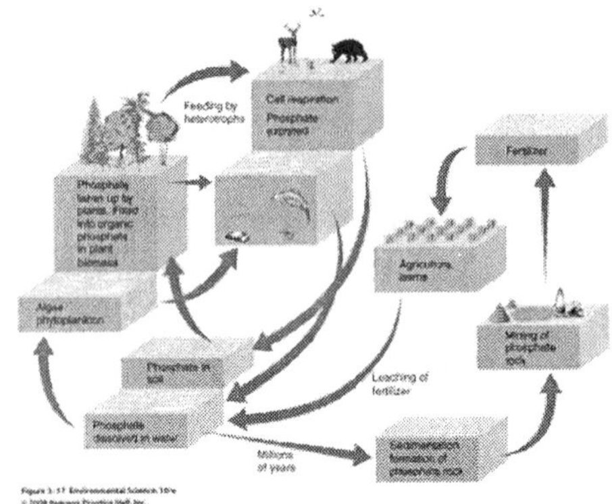

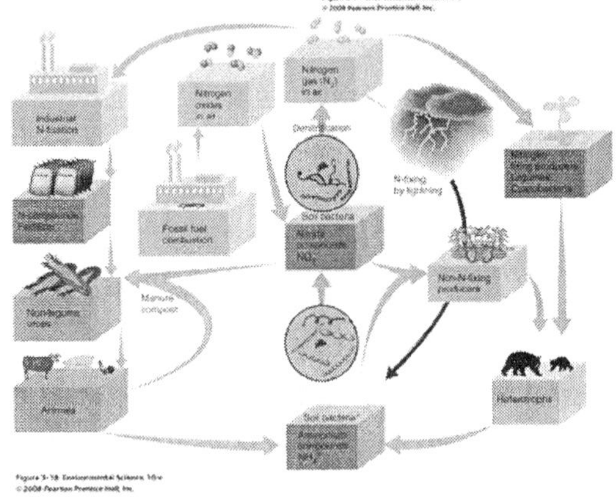

2. The law of conservation of matter states that atoms cannot be created, changed, or destroyed; they only can be rearranged. This law is demonstrated by carbon as it is moved or "fluxed" around the system through the carbon cycle.

Ecosystems: How They Change

4.1 Dynamics of Natural Populations

Figure 4.2

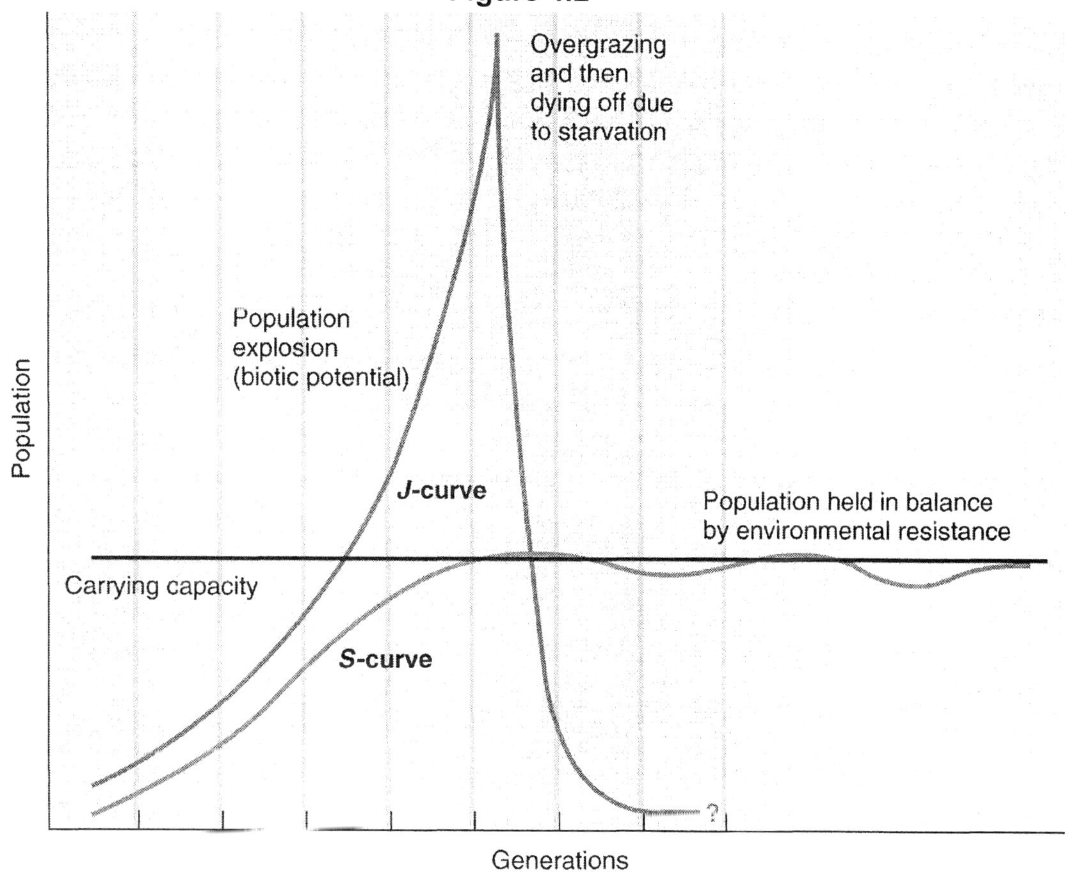

Figure 4-2 Environmental Science, 10/e
© 2008 Pearson Prentice Hall, Inc.

Exponential growth: Occurs when the population growth rate increases over time. When graphed, exponential growth looks like the letter "J" (Figure 4–2).

S-curve: After an explosive or exponential growth occurs, natural mechanisms take over to level off a population. When graphed, this population pattern results in a line that looks like an "S" (Figure 4–2)

Reverse J curve: If resources run out following an exponential growth in population, then the population experiences a huge die off. When graphed, this population pattern results in a line that looks like a backward or reverse "J" (Figure 4–2)

Carrying capacity is the maximum population that an ecosystem can support without being degraded (Figure 4-2).

Reproductive Strategies

- **r strategists** produce quickly and produce many offspring, but few reach adulthood. Frogs are an example of r strategists.

- **k strategists** produce more slowly and produce few offspring, but many reach adulthood. Humans are an example of k strategists.

- Low recruitment is another way to say "high mortality of the young."

- High recruitment is another way to say "low mortality of the young."

Biotic Potential	The number of offspring that a species may produce under ideal conditions
Environmental Resistance	The combination of all the biotic and abiotic factors that limit a population, such as loss of habitat, pollution or food resources
Replacement-Level Fertility	When animals have just enough offspring to replace the parents maintaining the population
Carrying Capacity	The maximum population that a habitat can support without the habitat being degraded over time

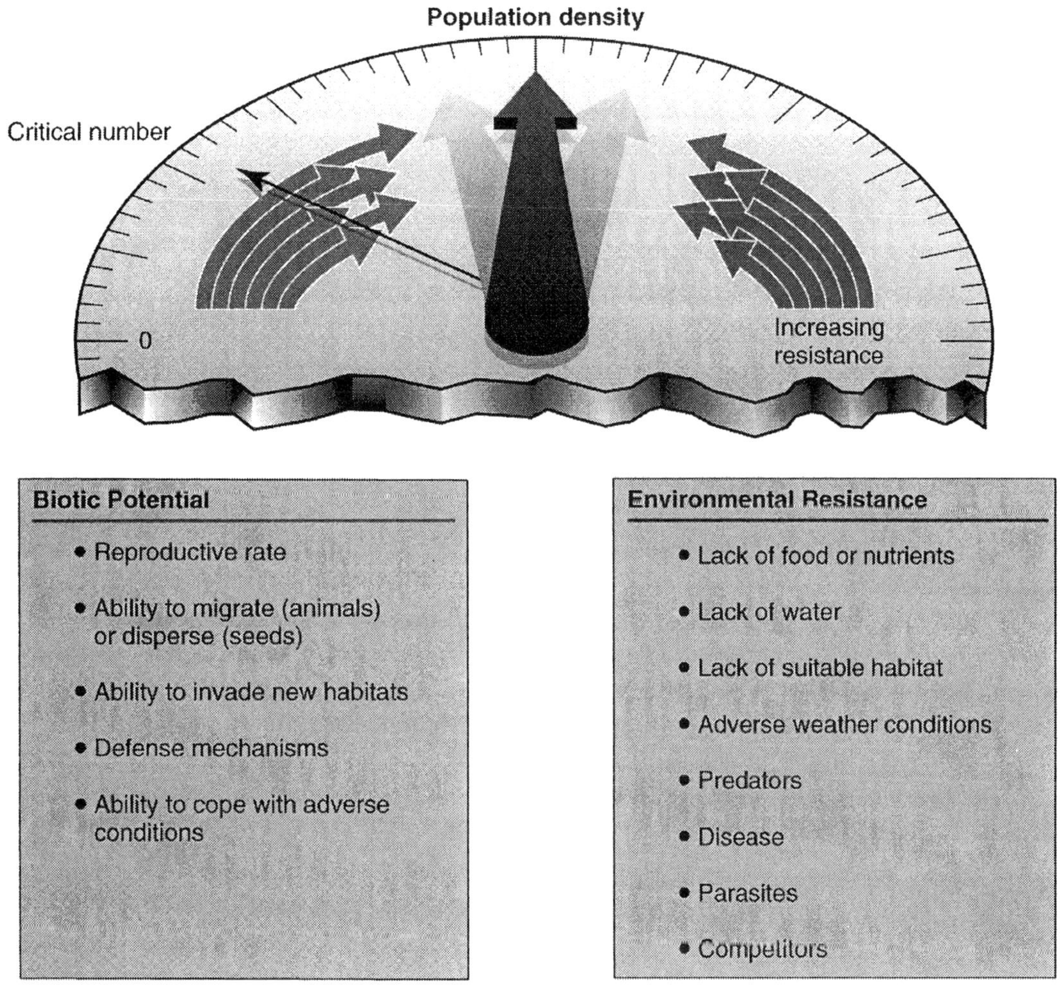

Figure 4-4 Environmental Science, 10/e
© 2008 Pearson Prentice Hall, Inc.

Figure 4-4

Density Dependent Factors	When the number of individuals becomes so great that the environment cannot support it and the population ceases or declines
Density Independent Factors	When a population ceases or declines for reasons other than the number of individuals (e.g., deep freeze)
Critical Number	The minimum population base that a group of individuals can recover from before it ceases or declines

4.2 Mechanisms of Population Equilibrium

Predator-Prey Dynamics

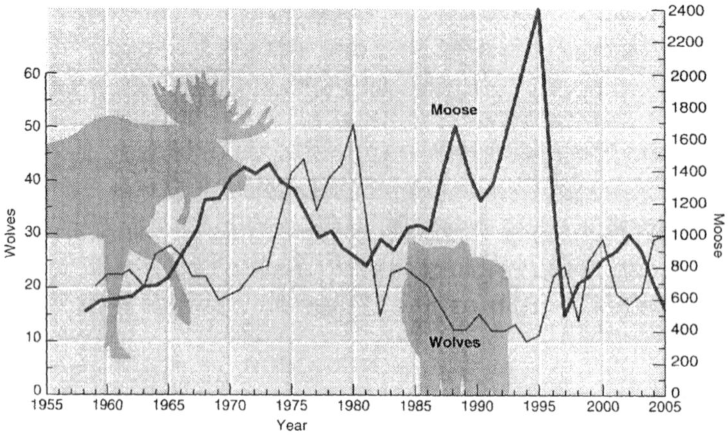

Figure 4-5 Environmental Science, 10/e
© 2008 Pearson Prentice Hall, Inc.

Predator-prey relationships are important to the health of each species, as illustrated in the example between wolves and moose in Figure 4–5. As the number of moose increases, so also do the number of wolves who prey on the moose. Consequently, a rise in the wolf population is followed by a decline in the moose population, which is itself followed by a decline in the wolf population. This up-and-down pattern will continue as long as there is a healthy predator-prey relationship.

Predators usually kill the sick, weak, or injured. This strengthens the rest of the herd that is preyed upon.

Parasites act as a type of predator by keeping the population in check in a density dependentmanner. If the population density of the host increases, parasites find new hosts. As a result, infection rates and mortality rates increase.

Herbivores are considered predators of plant species. This kind of predation can become a problem if an area becomes overgrazed.

Keystone species are animals that play an important role in maintaining an ecosystem. If a keystone species is removed, then other species in the ecosystem are greatly reduced.

Competition

Intraspecific competition is competition between members of the *same* species, and **interspecific competition** is competition between members of *different* species.

Introduced species (also called nonnative species) are animals or plants brought in to an ecosystem from a foreign ecosystem. Usually, an introduced species has no natural predators and thus is able to outcompete native species for space, food, or other resources. Some introduced species may take over an area. Rabbits overgrazing in Australia is a prime example.

4.3 Evolution as a Force for Change

Natural Selection

Natural selection are natural pressures (also called selective pressures) from the environment that determine a species survival. As a species faces these selective pressures, it has three choices: adapt, migrate, or become extinct.

There are four things that affect a species' survival:

- its geographic distribution;
- its specialization to a given habitat or food supply;
- its genetic variation within the gene pool of the species;
- its reproductive rate compared to the rate of environmental change.

Currently, the loss of biodiversity due to human impacts on the environment is one of the most devastating changes affecting different species' survival.

Speciation is the process by which populations of one species separate and develop into new species. For speciation to occur,

- the original population must separate into smaller populations that do not interbreed (**reproductive isolation**);
- the separated populations must experience different selective pressures.

The different finches that Darwin encountered on his trip to the different islands of the Galápagos are an example of speciation.

Drifting Continents

Plate tectonics is a theory that helps to explain volcanoes and earthquakes. It also helps explain why species are distributed as they are around the Earth.

The fossil record provides evidence that the Earth's continents used to be fused together. This "supercontinent" was named Pangaea.

Within the earth, molten materials rise to the surface and spread out. These rising and falling motions are called **convection currents**. As they float on top of these currents, crustal plates slowly move apart (about 6 centimeters a year), slide next to one another, or come together. In areas where plates come together (convergent boundaries), mountains or volcanic islands are formed. In areas where plates move apart from each other (divergent boundaries), mid-ocean ridges form. Faults and earthquakes occur in areas where plates slide past one another (transform boundaries).

4.4 Ecosystem Responses to Disturbance

Ecological succession is the change in an ecosystem from one community to another. The first species to move into an area is called the pioneer species, and the final species (which tends to be longer lived) is called the climax community.

Primary succession occurs when an area begins as bare rock and certain species of moss or lichen that are adapted to the environment cling to the rock. These species form a mat that can catch soil particles as they blow or wash by. Thus, a layer of soil, held in place by the moss, accumulates. This mat provides a place for seeds to lodge and larger plants to grow.

Secondary succession occurs after an area has been disturbed, such as by fire or humans, left alone, and then reinvaded by plants. Although it occurs after a series of stages, secondary succession happens faster than primary succession because the soil is already there.

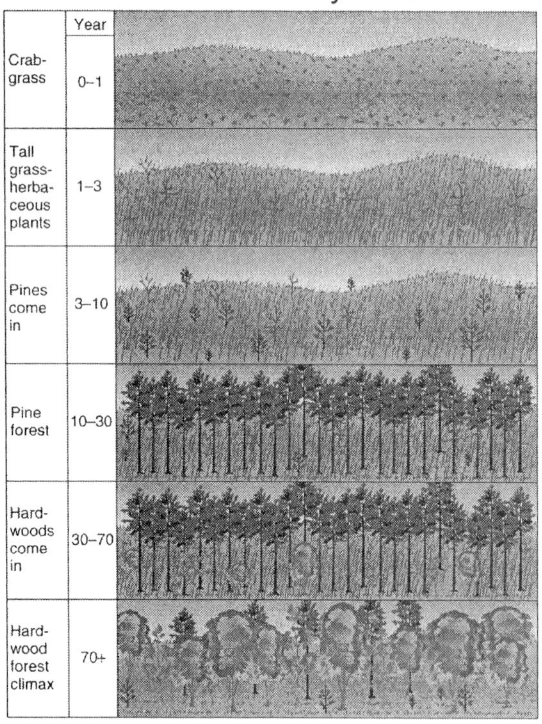

Figure 4-21a Environmental Science, 10/e
© 2008 Pearson Prentice Hall, Inc.

Figure 4-20

Succession also can occur in lakes or ponds, as soil settles and gradually fills them. A climax community can be a bog or a forest.

Multiple Choice Questions

1. In an aquatic ecosystem undergoing thousands of years of succession, the climax community will most likely be a

   A. deep lake
   B. bog or forest
   C. bare-bottom lake
   D. taiga
   E. bare rock

Use the graph below to answer question 2.

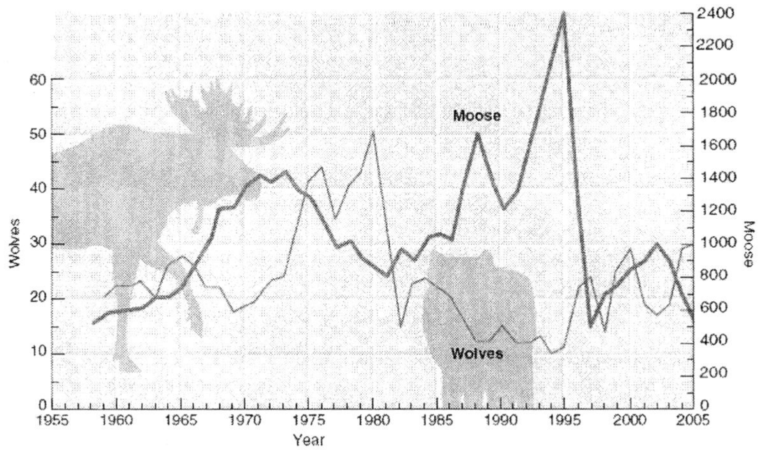

Figure 4–5 Predator–prey relationship. Wolf and moose populations on Isle Royale from 1955 to 2005.

2. What might the graph change if white tailed deer were introduced into Isle Royale?

   A. The wolf population might increase initially because of a decrease in environmental resistance.
   B. The moose population might decrease because of an increase in environmental resistance.
   C. The moose population might increase because of a decrease in environmental resistance.
   D. The moose population might decrease because of a decrease in biotic potential.
   E. A, B, and C

3. When will parasites have the least effect on prey population?

   A. When the prey population density is low, the predator population density is high.
   B. When the prey population density is high, the predator population density is high.
   C. When the prey population density is high, the predator population density is low.
   D. When the prey population density is low, the predator population density is low.
   E. When the prey population density is high, the predator is not present in ecosystem.

4. What occurs when a foreign species is introduced into an ecosystem?

   A. a definite long-run increase in biodiversity
   B. a definite long-run decrease in biodiversity
   C. a new, dynamic equilibrium in the ecosystem
   D. a fragmented edge effect on an ecosystem
   E. none of the above

Refer to the list below to answer questions 5–8.

A. Housefly, which lay many eggs and then abandon their nest
B. African elephants, which bear one calf and protect it for a few years

5. Which best illustrates a k strategist?

6. Which best illustrates an r strategist?

7. Which has the highest level of recruitment?

8. Which has the lowest level of recruitment?

Refer to the following graph to answer the questions 9–12.

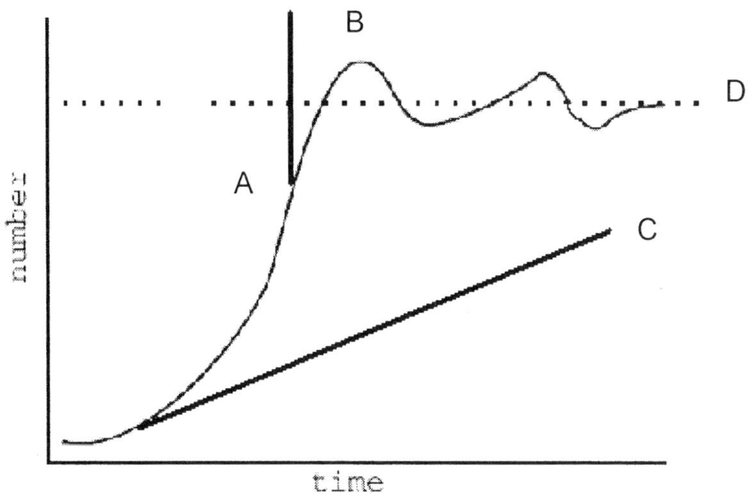

9. Which letter on the graph represents the number of individuals that the area can support for a long period of time?

10. Which letter on the graph represents exponential growth?

11. Which letter on the graph represents linear growth?

12. Which letter on the graph represents the growth that overshoots the carrying capacity?

Free Response Questions

Nonnative species can have a huge impact on the environment into which they are introduced. Cane toads in Australia are a famous example of how a nonnative species can take over.

Cane toads were brought to Australia from Hawaii, where the toads were being used successfully to help fight the cane beetles that were destroying sugar cane crops. Unfortunately, Australians soon found out the toads did not jump high enough to eat the cane beetles that stayed on the upper stalks of the cane plants. When the beetles' larvae were emerging from the ground, no toads were around. Consequently, the toads had no impact on the cane beetle problem in Australia, and farmers returned to using chemical pesticides to control the beetles.

Meanwhile, the toads breed very quickly, their tadpoles outcompete native frog tadpoles, and the cane toads, which are poisonous, have no natural predators. In addition, while the fish that eat cane-toad tadpoles die and the animals that eat cane toads die, many cane toads eat native Australian frogs with no dire consequences to themselves.

The cane toad's ability to breed very quickly and its lack of natural predators have led to a sizable population of cane toads in Australia. They have spread across most of Queensland and the Northern Territory, including the wetlands of Kakadu. Each pair of cane toads can lay 33,000 eggs per spawning. They are also able to eat almost anything, including dog food.

Cane toads have proven to be one of Australia's worst environmental disasters. These toads are responsible for the reduction of many species of Australian wildlife and have been very hard to control.

1. Using the graph paper below, draw the growth curve of the cane toads as they were introduced into Australia.

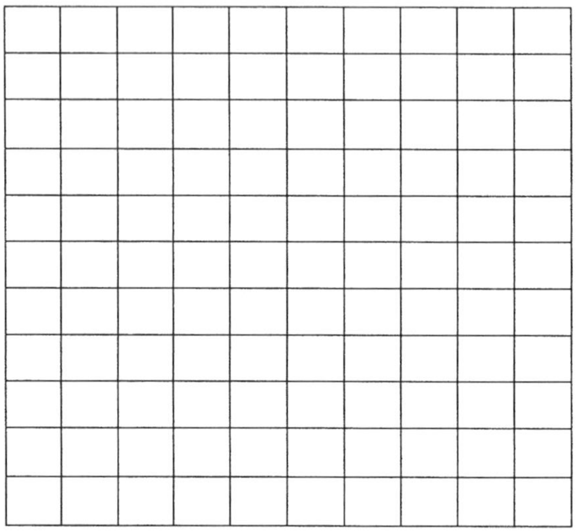

2. Explain the reproductive strategy of the cane toads. Include a discussion of the biotic potential and environmental resistance of the cane toads.

3. What has happened to the population density of the cane toads and why?

4. Discuss the predator-prey dynamics in the story of the cane toads.

Multiple Choice Answers and Explanations

1. B. Succession can occur in lakes or ponds as soil settles and gradually fills them. A climax community can be a bog or a forest.

2. E. The reintroduction of the white-tailed deer could decrease environmental resistance for the wolf because the deer represents a new food source. The deer might increase the environmental resistance for the moose because the deer might compete with the moose for food and/or habitat. The deer might reduce environmental resistance for the moose because the wolf would have an alternative food source, therefore putting less pressure on the moose population.

3. D. When the population density of both the prey and predator are low, parasites have fewer animals to infect.

4. C. The introduction of nonnative species into an ecosystem results in three outcomes: the species dies out, the species becomes part of the natural ecosystem, or (in the worst case) the species becomes invasive and outcompetes native species.

5. B. Since they have one offspring at a time and take care of the offspring until it can take care of itself, African elephants are k strategists.

6. A. Since they lay many eggs at one time and leave the offspring to fend for themselves, houseflies are r strategists.

7. B. The African elephants' offspring have a very good chance of surviving to adulthood, so they have the highest level of recruitment.

8. A. The green sea turtles' offspring have a poor chance of surviving to adulthood, so they have the lowest level of recruitment.

9. D. The dotted line labeled "D" is the carrying capacity, or the number of individuals the area can support for a long period of time.

10. A. Exponential growth is represented by the part of the line labeled "A." When graphed, exponential growth looks like the letter "J".

11. C. Linear growth is represented by the line labeled "C." Linear growth occurs when a fixed amount is added each time.

12. B. After exponential growth takes place, the population will overshoot the carrying capacity. This growth is represented by the part of the line labeled "B."

Solutions to Free-Response Questions:

1. See graph below.

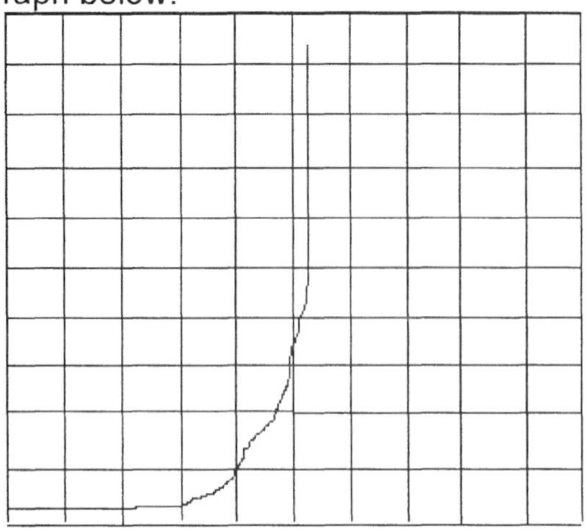

2. Cane toads are r-strategists. However, unlike most other r strategists, many of their offspring reach adulthood. This possibility shows the danger of nonnative species. The biotic potential of the cane toad is very high because of its high reproductive rate, its ability to migrate and invade new habitats, its poison as a defense mechanism, and its ability to cope with adverse conditions. The environmental resistance of the cane toad is low because it eats almost anything, has no natural predators, and has no competition in the environment.

3. The population density of cane toads is very high due to a high biotic potential and a low environmental resistance.

4. The predator-prey dynamics in the cane toad story illustrates one of the problems associated with introduced species. Since there are no natural predators in Australia to help keep the cane toad numbers in check, the cane toad population has been able to grow exponentially.

The Human Population

5.1 Human Population Expansion and Its Cause

The human population started rapidly, but has expanded significantly only in the last 100 years. Before, many fatal diseases such as smallpox, diphtheria, measles, and scarlet fever kept the human population in check. These diseases mainly affected infants and children. Although birth rates were high so were death rates; consequently, the population did not grow. Vaccinations were developed in the late 1800s, and penicillin was discovered in the 1930s. These two medical developments reduced the mortality rate. Since birth rates remained high, the population began growing almost exponentially. This population growth is expected to level off at about 10 billion by the year 2100. These questions remain:

- Where will all these people live?
- How will all these people be fed, clothed, housed, and educated?
- Will enough energy and material resources be available for all these people to enjoy a good life?
- What will the natural environment look like by 2100?

Important Terms You Need to Know

Growth Rate (annual rate of increase)	The percentage change in a population
Total Fertility Rate	The average number of children a woman has in her lifetime
Replacement-Level Fertility	The number of children a woman must have to replace her and her partner The number should be 2.0. However, because of infant and childhood mortality, it is slightly higher (2.1 for developed countries and higher in developing countries).
Infant Mortality	The number of infant deaths per 1,000 live births per year
Population Profile (age structure or histogram)	A bar graph showing the number of males and females at each age level for a country
Crude Birthrate	The number of live births per 1,000 people per year
Crude Death Rate	The number of deaths per 1,000 people per year
Doubling Time	The time it takes for a population to double in size
Demographic Transition	The tendency of a population to shift from high birth and death rates to low birth and death rates The result of a demographic transition is a population that grows very slowly or not at all.

5.2 Different Worlds

The World Bank divides countries into three economic categories.

High income, highly developed, industrialized countries (Developed countries)	United States, Canada, Japan, South Korea, Australia, New Zealand, Western Europe, Scandinavia, Singapore, Taiwan, Israel, several Arab states
Middle-income, moderately developed countries (Developing countries)	Mexico, Central America, South America, northern and southern Africa, China, Indonesia and other southeastern Asian countries, many Arab states, eastern Europe and countries of the former USSR
Low-income, developing countries (Developing countries)	Eastern, western, and central Africa, India and other countries of central Asia, and a few former Soviet republics

While the highly developed countries make up 15% of the world, they earn 80% of the world's income. The low-income, developing countries make up 37% of the world's population, yet they earn only 3.0% of the world's gross national income. That is a disparity of 63 to 1!

The developed world is growing at a rate of 0.1% a year, adding less than 1 million people a year. The developing world, in contrast, is growing at a rate of 1.5% a year, adding over 75 million people a year. (Ninety-eight percent of world population growth is occurring in developing countries.)

Paul Ehrlich and John Holdren came up with the following environmental impact equation:

$$I = P \bullet A \bullet T$$

This equation, referred to as the *IPAT formula*, states that environmental impact (I) equals the size of a population (P) multiplied by the country's affluence (wealth) (A) and by the country's level of technology (T). A high level of technology and a high level of affluence mean that developed countries have a large environmental impact, even though most have a fairly small population. For example, the average American places at least 20 times the demand on the earth's resources than does a person in a developing country.

5.3 Consequences of Population Growth and Affluence

In the past, humans grew their own food, lived on the land, and used only the resources they needed. With a small population, this lifestyle is sustainable. However, after World War II, advances in medicine caused death rates to fall and populations to grow rapidly. With a rapidly growing agricultural society, six options are possible:

- land reform, or redistributing land from the wealthy to the poor to increase equity.
- intensifying cultivation, or working the land harder, for example, by not rotating crops to allow the land to regenerate.
- opening up new land for agriculture, or converting natural landscapes into land for farming or ranching, for example, by cutting down forests to grow crops.
- engaging in illicit activities. When people are in poverty, they will break the law sometimes to survive.
- emigrating or immigrating to countries in search of a better life or to escape wars and ethnic persecution.
- migrating to cities in search of employment and a better life. Most rapidly growing cities experience poor sanitation and drinking water and rampant crime.

Most of the deprivation of poverty falls on the women and children of a population.

The United States leads the world in the consumption of many resources like fossil fuels and generates 24% of the world's carbon dioxide emissions. However, it also has many amenities that poor countries do not, including safe drinking water, sewage treatment, and waste disposal. Thus, many forms of pollution are held in check, and the environment improves with increasing affluence.

5.4 Dynamics of Population Growth

A **population profile** (also referred to as an age structure pyramid or histogram) is a bar graph that shows the number of people (males and females separately) at each age for a given population. Figure 5.13a is a population profile for Italy. It shows that Italy's population is getting smaller because there are less young people (aged 24 and below) than there are adults (aged 25 to 49).

Figure 5.13a

2005

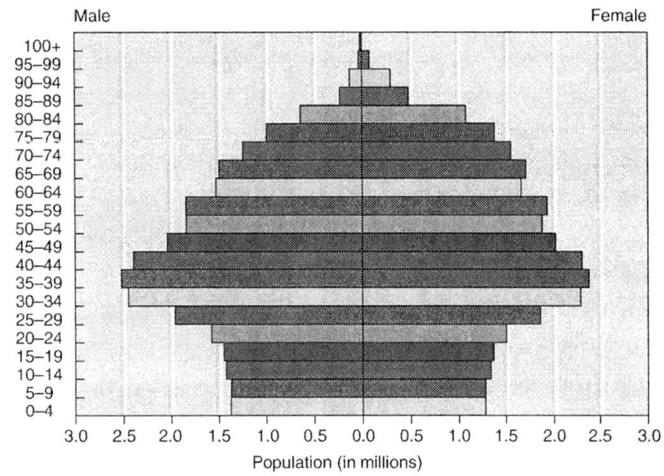

Figure 5-13a Environmental Science, 10/e
© 2008 Pearson Prentice Hall, Inc.

The population profile for the United States looks different than Italy's (Fig. 5.11b). The U.S. population is more stable than Italy's population. This stability is because the total fertility rate in the United States (2.0) is higher than the fertility rate in Italy (1.3).

Figure 5.11b

2005

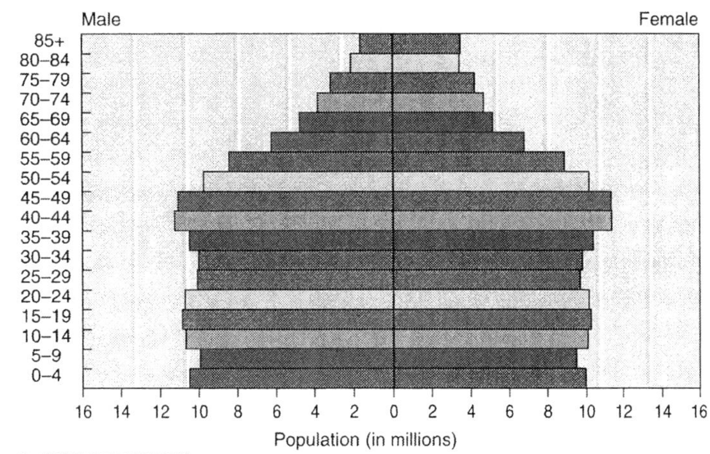

Figure 5-11b Environmental Science, 10/e
© 2008 Pearson Prentice Hall, Inc.

Demographers are people who study population trends and make projections about them. They can look at population profiles and make assumptions about fertility, mortality, and migration. The United Nations gives three different population projections: low, medium, and high.

Italy and other European countries face difficult economic decisions because of their declining populations. For example, who will produce the goods and services needed by their aging populations if the working-age population is declining? Will their economies remain competitive in the world market if their working-age population is small? Allowing more people to immigrate to these countries is one solution; however, many countries are hesitant about opening their borders, fearing such a move will lead to unwanted cultural and religious changes.

The situation is quite different in developing countries. Most developing countries' populations are soaring because of a large percentage of the population in their reproductive years. For example, the population profile shown in Figure 5.–15a is for Iraq. The broad-based pyramid shows that even a high infant mortality rate cannot offset the country's high fertility rate. Thus, the pyramid will get wider and wider.

Figure 5.15a

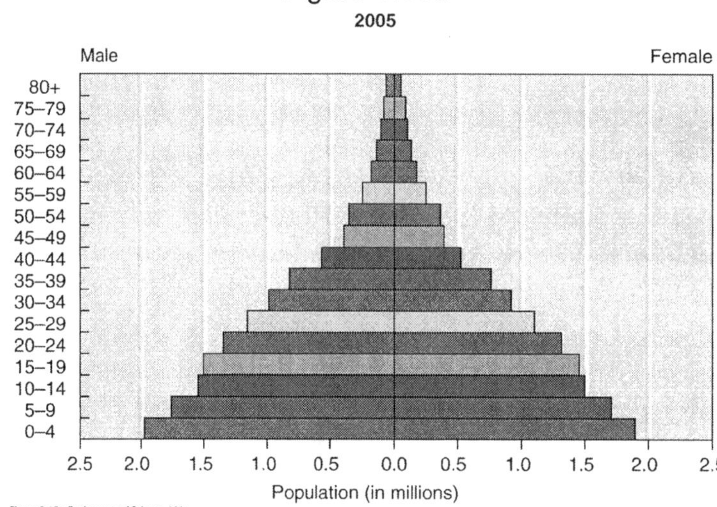

Figure 5-15a Environmental Science, 10/e
© 2008 Pearson Prentice Hall, Inc.

Iraq's population profile shows a phenomenon called **population momentum**. Iraq will continue to grow for 50 to 60 years, even after the total fertility rate is reduced to replacement level. The population is comprised of few older adults and many women entering their reproductive years. The number of births, therefore, will far exceed the number of deaths. Even if women have only two children per woman, the population will grow.

In contrast, Europe's population is projected to experience negative population momentum. This projection has lead many European countries to encourage people to have more children.

As countries industrialize, demographers have found that they go through a **demographic transition**, where high birth rates and high death rates are replaced by low birth rates and low death rates.
To calculate the **percent increase** (or decrease) per thousand per year of a country, the following equation is used:

(Crude birth rate (CBR) and crude death rate (CDR) are the number of births and deaths, respectively, per thousand of the population per year.)

Number of births per 1000 per year (CBR)	–	Number of deaths per 1,000 per year (CDR)	=	Natural increase (or decrease) in poulation	÷	10	=	Percent increase (or decrease) in population per year

The **doubling time**, or the number of years it will take for the population to double, is calculated by dividing the percentage rate of growth into 70.

The demographic transition is typically presented as four phases, as shown in Figure 5–18.

Figure 5-18

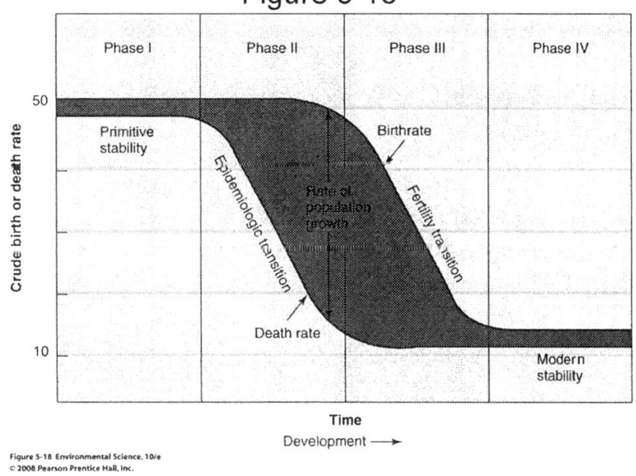

The following chart summarizes the characteristics of each phase:

Phase	Characteristics
I	A high CBR and a high CDR means the population does not grow.
II	A declining CDR due to medical advance along with continued high CBR means the population explodes.
III	Although the fertility rate declines as the country becomes industrialized, population growth is still significant.

Multiple Choice Questions

1. Changes to the size of a country's population depends on

   A. births
   B. deaths
   C. immigration
   D. emigration
   E. all of the above

2. The greatest environmental problem the world is facing today is

   A. pollution
   B. the human population explosion
   C. the greenhouse effect
   D. depletion of natural resources
   E. none of these

3. The crude birth rate is the number of births per __________ people per year.

   A. 50
   B. 100
   C. 500
   D. 1,000
   E. 10,000

4. Developed countries

   A. have a large replacement level fertility
   B. sizes are found by subtracting the number of immigrants from emigrants
   C. tend to consume less meat than developing countries
   D. generate the majority of the world's pollution and waste
   E. have high infant mortality rates

5. The environmental impact equation includes

   A. population size
   B. affluence
   C. technology
   D. all the above
   E. A & B only

6. The term *demographic transition* describes

   A. the movement of people from poor countries to wealthy countries
   B. the tendency of a country to move from high birth and high death rates to low birth and low death rates
   C. the time it takes for a population to double in size
   D. when all the veterans returned from World War II to start families
   E. all of the above

Use the following population profile to answer questions 7, 8, 9, and 10.

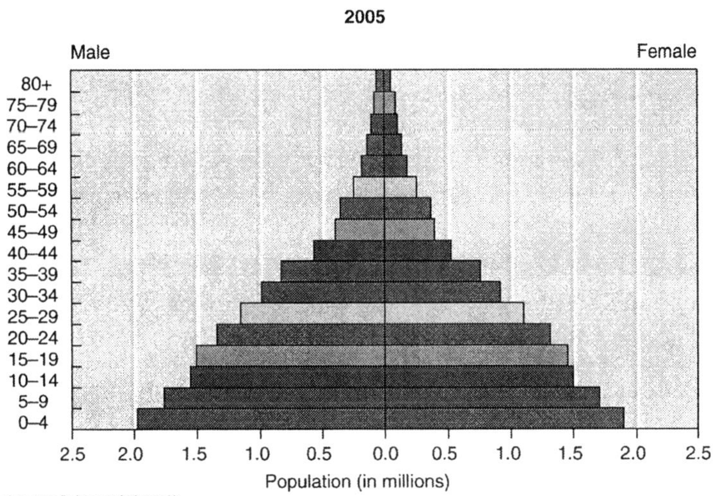

Figure 5-15a Environmental Science, 10/e
© 2008 Pearson Prentice Hall, Inc.

7. Which term applies to this population profile?

   A. developed
   B. developing
   C. moderately developed
   D. industrialized
   E. none of the above

8. This population is

   A. expanding
   B. stable
   C. getting smaller
   D. both A & B
   E. both B & C

9. The number of females aged 0 to 4 is approximately

   A. 1,900
   B. 4,000
   C. 2 million
   D. 3.8 million
   E. 1.9 million

10. Approximately what percent of the population is under age 20?

   A. 1%
   B. 10%
   C. 40%
   D. 75%
   E. 90%

Free Response Questions

1. If a population of 10,000 experiences 70 births, 40 deaths, 30 immigrants and 50 emigrants in a year, what is the net annual percentage growth rate?

2. If a country's population grows at a yearly rate of 3%, in how many years will the population double?

3. Compare the rates of population growth in developed countries and developing countries. Explain the differences you find.

4. What are some issues that countries face when their populations are shrinking?

Multiple Choice Answers and Explanations

1. E. To calculate a population, you take birth rate minus death rate plus immigration minus emigration and divide by 10.

2. B. The human population explosion affects all of the other environmental problems, making it the greatest environmental problem the world faces.

3. D. The crude birth rate is the number of live births per 1,000 people per year.

4. D. A high level of technology and a high level of affluence make developed countries have a large environmental impact, even though most have a fairly small population. For example, the average American places at least 20 times the demand on the earth's resources than does a person in a developing country.

5. D. The Environmental Impact Equation developed by Paul Ehrlich and John Holdren is I = P • A • T, or Environmental Impact equals the size of a population multiplied by the country's affluence (wealth) and how much technology it has.

6. B. Demographic transition is the tendency of a population to shift from high birth and death rates to low birth and death rates. The result is a population that grows very slowly or not at all.

7. B. Any population profile shaped like a pyramid is developing.

8. A. A population profile shaped like a pyramid means the size of the population is expanding.

9. E. There are approximately 1.9 million females aged 0 to 4, according to the population profile shown.

10. C. The bottom four bars of the population profile (males and females aged 0 to 19) account for approximately 40% of the entire population.

Free Response Answers

1. 70-40 + 30-50 / 10,000 X 100%
   30 + (-20)/10,000 X 100%
   10/10,000 X 100% = 0.1%

2. 70/3 = 23 years

3. Developed countries are growing slowly, at a rate of 0.1% a year. Developing countries are growing at a faster pace, around 1.5% a year. This difference in growth rates mean that 98% of the world's population growth is occurring in developing countries.

4. Issues facing countries with negative growth rates are usually economic. Who will produce the goods and services needed by an aging population, and will the country's economy remain competitive? These issues have led some European countries to encourage their people to have more children.

Population and Development

6.1 Reassessing the Demographic Transition

The eighteenth-century British economist Thomas Malthus observed that populations tended to grow exponentially whereas agricultural output only increased linearly. He believed this difference would result in famine, as the quickly increasing population exhausted the slowly increasing food supply. Malthus was unable to see that the industrial revolution would bring huge developments in agriculture and that fertility rates would decline. However, Malthus' general concern about the human population is still relevant today.

At the **International Conference on Population and Development** (ICPD) in Cairo, Egypt, in 1994, issues of poverty, population growth, and development were discussed. Participants of the conference reached broad agreement on the following points:

- development must be linked to a reduction in poverty;
- the existing poverty in the developing countries is an affront to human dignity;
- both poverty and development are a threat to the health of the environment.

The following chart lists and describes several reasons why poor people in developing countries have a lot of children:

Security in one's old age	Elderly people are cared for by their children.
Infant and childhood mortality	When infant and childhood mortality is high, people have more children to ensure some make it to adulthood.
Helping hands	Children can work around the house or in the fields and bring money to the family.
Lack of importance of education	Most children (especially girls) in the developing world are not educated. If they are educated, they do not work and become a liability.
Status of women	If women have no education or career, the only opportunity they have is to bear children.
Availability of contraception	When contraceptives are unavailable or too expensive, fertility rates remain high.

Many women in developing countries state that they want fewer children or at least want to wait before having a second child.

Industrialized nations have the following factors that are conducive to having smaller families:

- a higher cost of raising children;
- existence of pensions and Social Security systems;
- more opportunities for women to join the workforce;
- easier access to inexpensive contraceptives;
- better health care;
- more inclusive educational opportunities;
- higher educational attainment;
- older age at marriage.

6.2 Promoting Development

In 1997, representatives from the United Nations, the World Bank, and the Organization for Economic Cooperation and Development met to set goals that would address poverty and its aspects. Below is a chart of the eight goals they set.

Eradicate extreme poverty and hunger
Achieve universal primary education
Promote gender equality and empower women
Reduce child mortality
Improve maternal health
Combat HIV/AIDS, malaria, and other diseases
Ensure environmental sustainability
Forge a global partnership for development

6.3 A New Direction: Social Modernization

In order to shift from high to low fertility rates, the following efforts must be made:

- **improve education**, especially literacy (learning to read, write, and do simple calculations). Because educating women is not seen as important, illiteracy rates among poor women are between 50 and 70%.

- **improve health**, including basic nutrition, hygiene (e.g., boiling water to prevent diseases like diarrhea), prenatal and postnatal care of mothers, reproductive health care, and education of young people. HIV/AIDS is the leading cause of death in sub-Saharan Africa.

- **make family planning accessible** so people can plan their own family size and when they want to have children. The IPCD concluded that abortion should never be used as a means of family planning. However, according to the World Health Organization, 46 million abortions occur each year, and more than one-third of these procedures are unsafe. (Approximately, 68,000 women die annually from unsafe abortions.)

- **enhance income through employment opportunities.** Many developing countries would like to provide goods or services, but do not have the money to do so. The Grameen Bank engages in microlending, or small, short-term loans (average $67) that can be used to purchase the basic things needed to start a business.

- **improve resource management,** which means helping to train people to sustain their resources through better management.

6.4 The Cairo Conference

At the conference in Cairo, all 179 nations agreed that population is an issue of crisis proportions that must be confronted. These nations signed the Program of Action, indicating their commitment to

- empower women;
- meet people's needs for education and health;
- advance gender equality;
- eliminate violence against women;
- enable women to control their own fertility.

Multiple Choice Questions

1. All of the following are reasons a family in a developing country would have more children *except*

   A. high infant mortality rate.
   B. more children means more help around the house or farm.
   C. more children means they take care of their parents in their old age.
   D. more children indicates higher educational attainment and a reduced level of poverty.
   E. more children means a lower poverty threshold for the family.

2. All of the following are social determinates that decrease birth and fertility rates *except*

   A. availability of reliable birth control.
   B. access to a clean and abundant water supply.
   C. urbanization.
   D. religious beliefs.
   E. increased literacy level.

3. International relief organizations may distribute relief aid to women through schools. This encourages all of the following *except*

   A. higher educational attainment for women.
   B. a decrease in malnutrition.
   C. higher fertility rates.
   D. eliminating violence against women.
   E. a higher incidence of birth control.

4. Sub-Saharan Africa is at a crossroads because of which of the following diseases?

   A. HIV/AIDS
   B. malaria
   C. influenza
   D. small pox
   E. all of the above

5. Thomas Malthus argued that

    i. population increases exponentially.
    ii. food production increases exponentially.
    iii. population increases linearly.
    iv. food production increases linearly.

    A. i and ii
    B. iii and iv
    C. ii and iii
    D. i and iv
    E. none of the above

Free Response Questions

1. Many demographers are concerned about the rate human population growth.

    a. Explain why women are key to solving this problem.
    b. What solutions were promoted at the 1994 conference in Cairo?
    c. Why would addressing poverty affect human population growth?

Multiple Choice Answers and Explanations

1. D. Most children in developing countries do not go to school because they are needed around the house or farm to help with the family chores.

2. D. Religious beliefs almost always increase birth rates.

3. C. When women are educated, fertility rates fall.

4. A. HIV/AIDS is at epidemic proportions in sub-Saharan Africa.

5. D. Thomas Malthus argued population grows exponentially and food production grows linearly. This trend, he believed, eventually would lead to starvation, disease, crime, and corruption.

Free Response Answers

1a. Studies have shown that the number one way to solve the human population problem is to target women. Women in developing countries tend to be undereducated and underemployed, leaving them only the option to bear children. When given the option, most women state that they would like to have fewer children.

1b. According to the conference in Cairo in 1994, the keys to stabilizing human population growth are to

- empower women;
- meeting people's needs for education and health;
- advance gender equality;
- eliminate violence against women;
- enable women to control their own fertility.

1c. In the developing world, poverty leads to high fertility rates. Many families cannot afford birth control, even if they want it. Children are considered assets to poor families because they can work to provide food or money for the family. Even in developed countries, high fertility rates can be linked to poverty.

Water: Hydrologic Cycle and Human Use

One of the world's worst environmental disasters is the Aral Sea. The Aral Sea is located in Kazakhstan and Uzbekistan, which were formerly part of the Soviet Union. The Aral Sea was once a huge salt lake with two fresh water feeder rivers. The Aral Sea's two large rivers have been used to irrigate millions of hectares of land. However, the draining of the Aral Sea has had a devastating effect on the ecology of the area. The lake is now three times as salty as the ocean, which has resulted in the death of many fish that once lived in the lake and the collapse of the local fishing industry. Today, the lake has 90% less volume and is 53 feet lower than it was in 1960. Salt from the lake blows onto crops, which makes the farmers pump more fresh water out of the rivers to flush out the salt. This has caused a positive feedback loop, and the situation is getting worse.

7.1 Water: A Vital Resource

Figure 7.1 provides some facts and statistics about the Earth's water.

- 71% of the Earth is covered by water.
- 97.5% of this surface water is salt water.
- Of the 2.5% that is fresh water, two-thirds is in polar ice caps and glaciers.
- Only 0.77% of all water is in lakes, rivers, groundwater, etc.

Figure 7.1

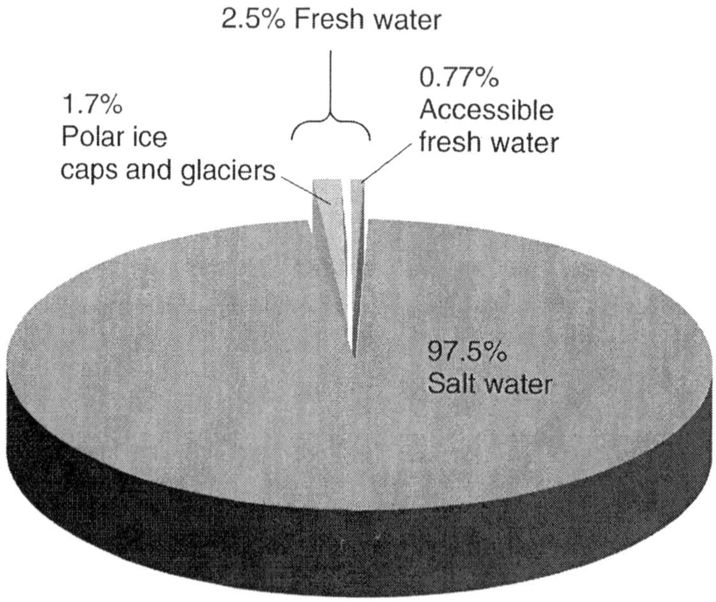

Figure 7-1 Environmental Science, 10/e
© 2008 Pearson Prentice Hall, Inc.

7.2 Hydrologic Cycle: Natural Cycle, Human Impacts

The **hydrologic cycle** (Figure 7–3) is how water moves through the atmosphere, land, and oceans.

Figure 7-3

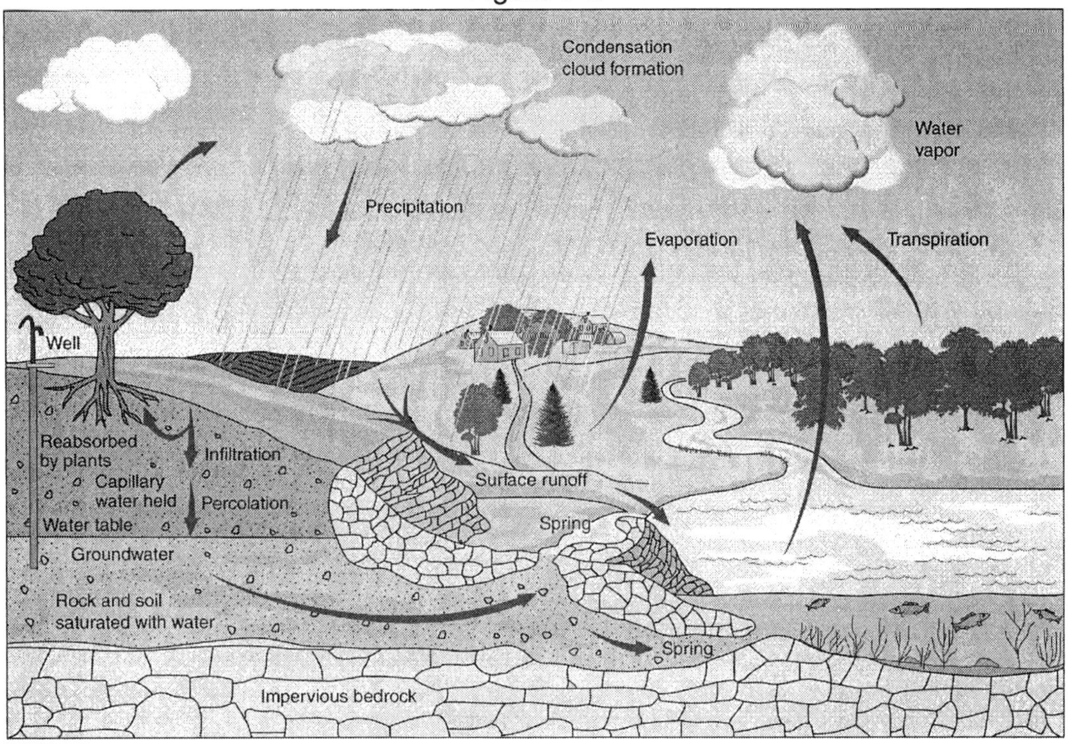

Figure 7-3 Environmental Science, 10/e
© 2008 Pearson Prentice Hall, Inc.

Precipitation, in the form of rainfall, sleet, hail, or snow, occurs when condensation intensifies.

Evaporation is when water leaves the liquid state and becomes a gas.

Condensation, the opposite of evaporation, occurs when water goes from a gas back to a liquid.

Transpiration is the loss of water from plants.

Precipitation

Warm, moist air rises because it is less dense than the colder air above it. The higher altitude has less pressure and is colder, causing the air to cool and expand, a process known as **adiabatic cooling**. As the air cools, it is unable to hold as much moisture, and precipitation occurs. **Adiabatic warming** occurs as the now cool air descends (because it is more dense) and becomes subject to the lower atmosphere's higher pressure.

This process occurs globally because of **convection currents** and more locally because of the **rain shadow** effect.

Convection currents are the vertical movement of air due to atmospheric heating and cooling.

Hadley cells, comprised of the two processes of adiabatic cooling and adiabatic warming, create major weather patterns, as shown in Figure 7–6a and b.

Figure 7-6a and b

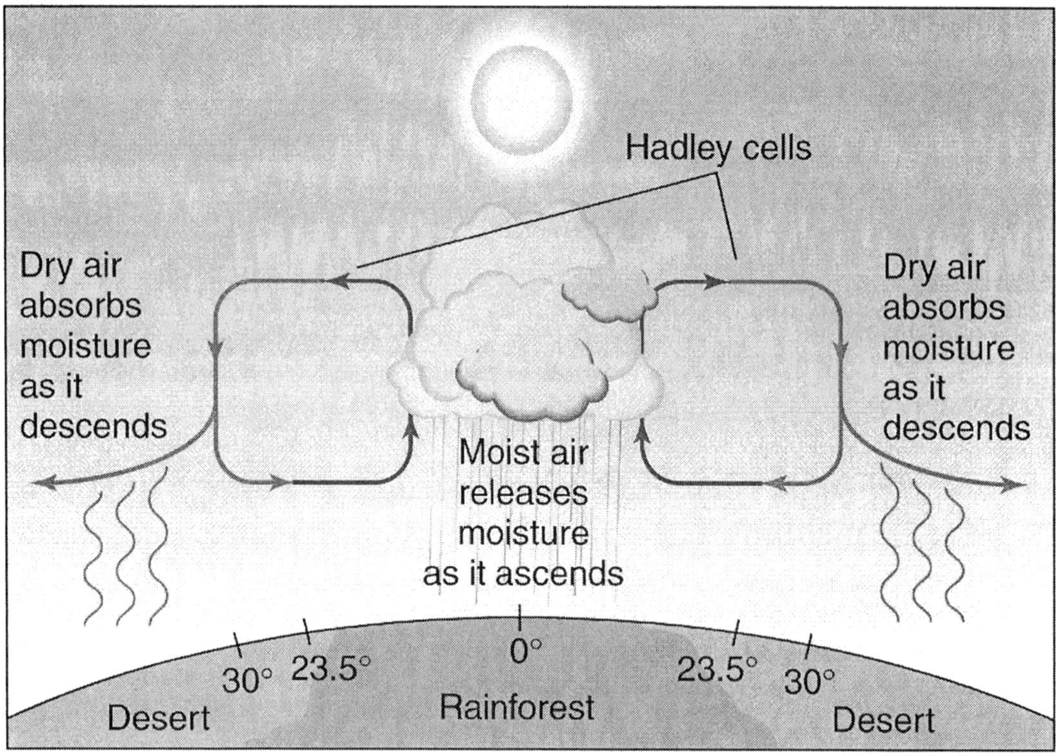

Hadley cells at the equator

Figure 7-6a Environmental Science, 10/e
© 2008 Pearson Prentice Hall, Inc.

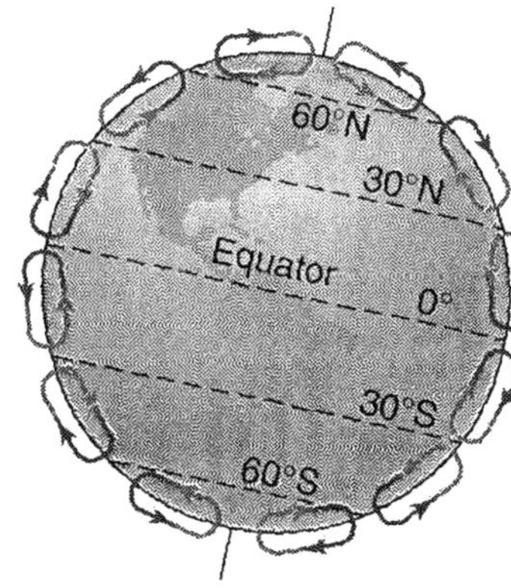

Global air flow patterns

Figure 7-6b Environmental Science, 10/e
© 2008 Pearson Prentice Hall, Inc.

As shown in Figure 7–7, the rain shadow effect occurs when moisture-rich air encounters a mountain range and is deflected up the mountain range. As the air moves up, it cools and expands, causing precipitation on the windward side of the mountain range. The now cool, dry air crosses the mountain range and descends, causing an arid climate on the opposite, or leeward, side of the mountain.

Figure 7-7

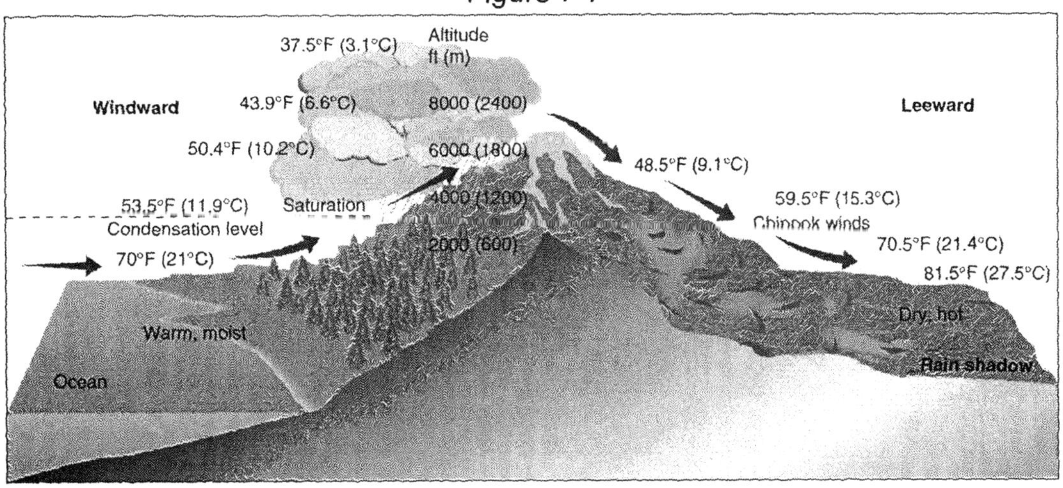

Figure 7-7 Environmental Science, 10/e
© 2008 Pearson Prentice Hall, Inc.

Precipitation either soaks into the ground and fills up aquifers, or it runs off into rivers and lakes and eventually ends up in the ocean. Precipitation that ends up soaking into the ground is called **groundwater**.

Today, due to deforestation and overgrazing, more water ends up as runoff than as groundwater. One consequence of more runoff is an increase in the number of floods. Increased runoff also means less evaporation, which mean less rainfall and drier conditions. Arid conditions, in turn, lead to more dust and smoke in the air, which forms a brownish haze that further suppresses rainfall.

7.3 Water: A Resource to Manage, a Threat to Control

The major global uses of freshwater are the following:

- 70% is used for irrigating crops.
- 20% is used for industry (power plants, etc.).
- 10% is used for direct human use (bathing, cooking, drinking, etc.)

An estimated 90% of all wastewater in developing countries is put directly back into lakes and rivers without any wastewater treatment. In many countries, this water is then consumed without being properly cleaned. The WHO estimates that 1.7 million people per year (mainly children) die from contaminated water.

Many rivers have been dammed to trap and store water. Some of the negative effects dams have on the environment include

- loss of waterfalls and rapids;
- loss of fish and other aquatic organisms such as salmon that migrate upstream to spawn;
- flooding of areas not previously underwater;
- draining of areas once aquatic;
- loss of biodiversity like waterfowl and wildlife that once utilized the river;
- the raising of salt concentrations in **estuaries** (places where the river meets the ocean) by allowing less freshwater to flow downstream .
- Loss of nutrient rich sediments (silt) downstream

Groundwater is also tapped to supply water for human use. The excessive withdrawal of groundwater has many problems associated with it. For example, the Ogallala aquifer, probably the largest aquifer in the world, is being drained so rapidly that its water table is dropping by about 6 feet per year. This aquifer is essentially nonrenewable since it is replenished only after many centuries. (The last time it was recharged was during the last ice age.) Other water tables are falling due to a similar overdrawing of groundwater. Falling water tables, in turn, affect surface water, including wetlands, streams and rivers. Furthermore, when water is overdrawn from aquifers, there is less support for the land above, which causes the land to sink and sinkholes to form. Along the coast, lowered water tables result in the intrusion of salt water into aquifers, as shown in Figure 7–18b.

Figure 7-18b

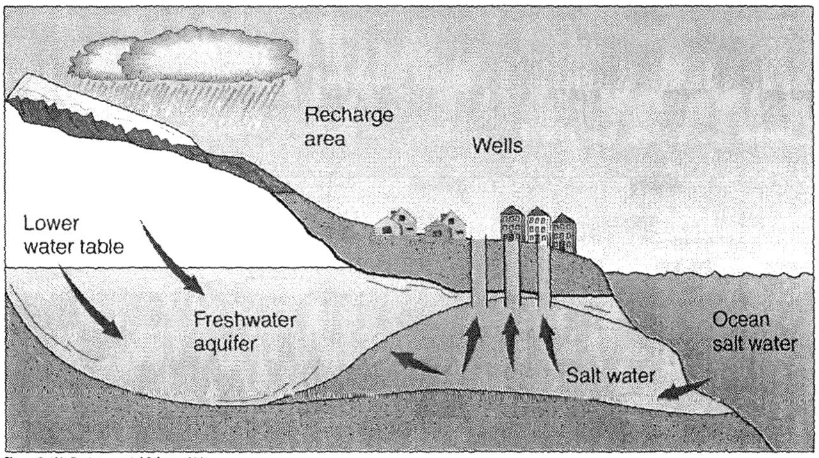

Figure 7-18b Environmental Science, 10/e
© 2008 Pearson Prentice Hall, Inc.

7.4 Water Stewardship: Public Policy Challenges

Currently, we have four options to meet the human population's growing demand for water in a sustainable way.

- We can build more dams to capture runoff water.
- We can gain better access to existing groundwater sources.
- We can desalinate seawater.
- We can conserve the water we have by using less of it.

Multiple Choice Questions

1. As more people move to cities, the amount of available groundwater is reduced. This is because of

A. more surface runoff.
B. greater evaporation.
C. greater transpiration.
D. higher humidity.
E. dust contributing to atmospheric pollution.

2. What would most likely happen to a forest that has been clear cut?

A. Biodiversity will increase.
B. The depth of soil will increase.
C. The water supply will be unaffected.
D. The amount of water in local rivers will increase.
E. Pollutants such as nitrates and phosphates will increase in the river.

3. The largest use of freshwater around the world is

A. watering people's yards.
B. doing laundry, showering, and other cleaning processes.
C. drinking water.
D. irrigating crops.
E. operating power plants.

4. As wetlands are drained due to the damming of rivers,

A. biodiversity is increased.
B. salt water concentration in wetlands is decreased.
C. freshwater concentration in wetlands is decreased.
D. the amount of sediment in wetlands is decreased.
E. the amount of pollutants in wetlands is decreased.

5. Which of the following would improve the Aral Sea's ecological problems?

A. Allowing farmers to continue pumping more water out of rivers to flush salts from the root zone of crops.
B. Teaching farmers to use drip irrigation and other water-saving techniques on crops.
C. Encouraging people living around the Aral Sea to have more children to help around the farms.
D. Damming the Amu Darya and Syr Darya rivers to control the flow of water downstream.
E. Having farmers pump salt water out of the sea rather than out of the rivers to put onto their crops.

6. The Ogallala aquifer is

A. a very small aquifer that is virtually nonexploitable.
B. an aquifer with a very fast recharge rate.
C. used to supply one-fifth of America's water to irrigate crops.
D. increasing by about 6 feet per year.
E. is predicted to be the country's only available drinking water by 2020.

7. Which of the following is true regarding desalinization?

A. Desalinization is very cheap.
B. Most countries around the world use desalinization for their drinking water.
C. Desalinization is a viable option to supply water for irrigation.
D. The greatest amount of desalination happens in the United States.
E. Reverse osmosis and distillation are commonly used for desalination.

8. The severest deserts in the world are caused by

A. overdrawing surface water.
B. overdrawing groundwater.
C. lack of technology to appropriately conserve water.
D. the rain shadow effect.
E. watersheds and surface waters.

9. Approximately what percentage of the Earth's water is liquid freshwater?

A. 97%
B. 50%
C. 32%
D. 2.5%
E. less than 1%

10. Hadley cells are a result of

A. rising air over the equator spilling over north and south at 30°.
B. infiltration of groundwater into an aquifer.
C. condensation of water molecules by hydrogen bonding.
D. surface runoff.
E. anthropogenic air pollution.

11. As air near the Earth's surface warms, all of the following occur *except*
A. air rises.
B. air expands.
C. adiabatic cooling occurs.
D. water vapor condenses.
E. air is deflected toward the equator.

12. Desertification in eastern California is primarily the result of

A. human impact on the environment.
B. global convection currents.
C. rain shadow effect.
D. aquifer depletion.
E. evapotranspiration.

Practice Free Response Question:

1. Explain how deforestation in the tropics would impact all of the following:

a. evapotranspiration
b. precipitation
c. surface runoff
d. water pollution
e. air pollution

Multiple Choice Answers and Explanations

1. A. As people cover soil with roads, buildings, etc., less precipitation is able to percolate downward into aquifers and more runs off into rivers and streams.

2. E. As forests are clear cut, pollutants such as nitrates and phosphates from the land run off into and contaminate rivers and streams.

3. D. Crop irrigation accounts for 70% of all freshwater use worldwide.

4. C. When rivers are dammed, the amount of freshwater entering estuaries is reduced and held back by the dam in the form of lakes and reservoirs.

5. B. Conservation methods such as drip irrigation has been proven to reduce water use by 30 to 70%.

6. C. One-fifth of all American crops are irrigated from water out of the Ogallala aquifer.

7. E. Reverse osmosis and distillation are commonly used for desalination.

8. D. The rain shadow effect is the low-rainfall region that exists on the leeward side of a mountain range.

9. E. Only .77% of all freshwater is found in lakes, wetlands, rivers, groundwater, biota, soil, and the atmosphere.

10. A. Hadley cells are two halves of a system composed of the rising and falling air that descends over subtropical regions (25 to 35 degrees north and south of the equator).

11. E. Air is deflected away from (not toward) the equator.

12. C. The rain shadow effect causes the severest deserts in the world, including Death Valley, which lies east of the southern Sierra Nevada mountain range in eastern California.

Free Response Answers

1a. Fewer plants mean less evapotranspiration, which could impact the area's climate by reducing the amount of water vapor in the atmosphere.

1b. If there was fewer plants, the air near the tropics would become hotter and drier and reduce the amount of moist air that would be available to cause precipitation.

1c. Deforestation would increase surface runoff because there would be less plants to help catch and hold precipitation. Consequently, less precipitation would infiltrate into groundwater and more would run off into rivers, streams, and lakes.

1d. The amount of pollutants in the area's water would increase due to more surface runoff to lakes, rivers, and streams carrying pollutants. Pollutants in the groundwater also would increase because there would be less plants available to filter out such toxins.

1e. Air pollutants would increase as the result of more dust and soil becoming airborne. Particles in the air form nuclei for condensation, and the more particles there are, the greater the tendency for clouds to form. These clouds would form a brownish haze that would suppress rainfall, causing drier conditions and additional dust and smoke.

Soil: Foundation for Land Ecosystems

8.1 Soil and Plants

Soil profiles refer to the horizontal layers of soil with distinct characteristics at each layer, as shown in Figure 8–4.

Figure 8-4

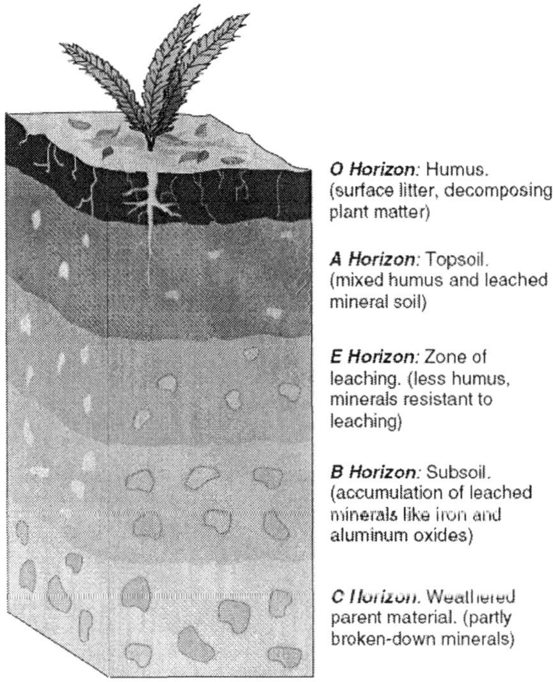

The **O horizon** is the topmost layer. It contains dead organic matter, such as leaves, stems, fruits, and seeds. The dark and rich soil that comprises this layer is formed from decomposition and is called humus.

The **A horizon**, also known as topsoil, is a mixture of the soil from below and the humus from above.

The **E horizon** is also referred to as the zone of eluviation. Eluviation is the process of dissolving away, or leaching, minerals as water moves through the layer.

The **B horizon**, also known as subsoil, has many minerals such as iron, aluminum and calcium that are leached from the layers above.

The **C horizon,** also known as parent rock, can be saturated in ground water.

Soil texture refers to the proportions of each of three particle types in soil. The different particles, classified by size, are

- Sand (particles from 2.0 to 0.02 mm in size)
- Silt (particles from 0.02 to 0.002 mm in size)
- Clay (particles less than 0.002 mm in size)

Touching is a good way to tell the difference between these three particle types. By touch, sand will feel gritty, silt will feel soft like flour, and clay will feel sticky and clumpy.

Loam is a mixture of all three soil types. It is approximately 40% sand, 40% silt, and 20% clay. Farmers consider loam to be a good soil type.

A soil texture triangle, like the one shown in Figure 8.3, can help determine what type of soil you have. Put soil in a test tube with water. The sand will sink to the bottom first because it is the heaviest. The silt will settle out later, and the clay will be suspended in the water the longest. Next, calculate what percentage of the soil in your test tube is each type. Then match up the percentages along each side of the soil texture triangle to determine your soil texture. Farmers use this technique to decide what crops will grow best and what they might need to add to the soil to maximize crop yields.

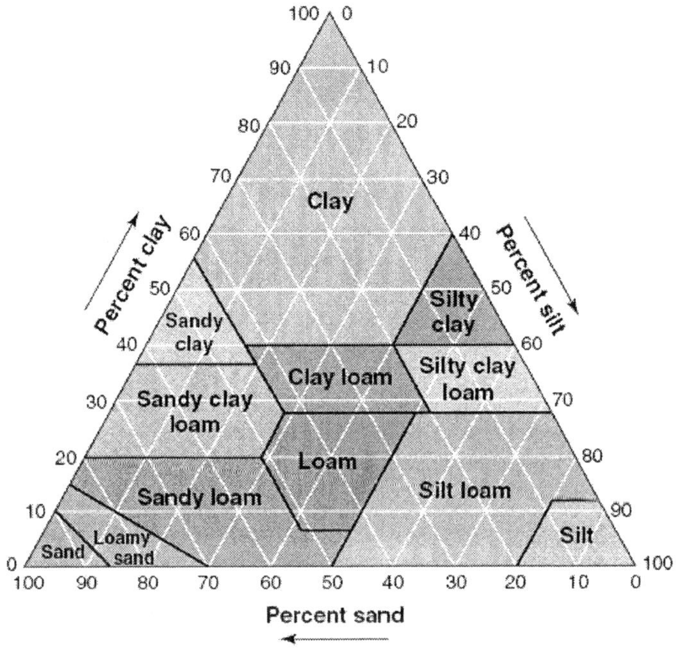

Figure 8–3 The soil texture triangle. Relative proportions of sand, clay, and silt are represented on each axis. Major classes of soil are indicated on the triangle. For clay, read across horizontally; for silt, read diagonally downward; for sand, read diagonally upwards to the left. The texture content of any soil should total 100% if the triangle is read properly.

Soils are classified by their structures and textures. The following four soil types are important to agriculture, animal husbandry, and forestry:

- **Mollisols** are fertile, dark soils found in temperate grassland biomes. They are considered the best agricultural soils.
- **Oxisols** are soils found in tropical and subtropical rain forests. Because most of the organic material is found in living plants, rain-forest soil is infertile.
- **Alfisols** are moderately weathered forest soils found in moist, temperate forest biomes. These soils are adequate for agriculture if supplemented with fertilizers or organic material.
- **Aridisols** are soils found in drylands and deserts. These soils are thin and light colored and usually contain a lot of sand. They are very susceptible to salinization.

Fertilizers can be either organic or inorganic. **Organic fertilizers** contain plant and animal waste, such as manure and compost. While they have high water-holding capacities, organic fertilizers do not have the exact amounts of the necessary nutrients farmers may want to add to their crops. **Inorganic fertilizers** are man-made chemical formulas. While they have the exact amounts of nutrients (like phosphates and nitrates that are limiting factors for plants) farmers need, they have poor water-holding capacities and leach from the soil very quickly.

A good soil for crops has

- a good supply of nutrients;
- ample infiltration and good water-holding capacity;
- good aeration;
- a pH near neutral;
- a low salt content

Soil structure refers to the arrangement of soil particles. A loose soil is ideal for infiltration, aeration, and workability.

8.2 Soil Degradation

Erosion is the process of soil and humus particles being picked up and carried away by water or wind. In natural ecosystems, plant cover and roots protect soil from erosion. However, when forests are clear cut or land is overgrazed, erosion can result.

Examples of water erosion are:

Splash erosion is when raindrops hit soil and remove it.

Sheet erosion is when a small layer of soil is removed from an entire area. Often, with this type of erosion, the results are not noticed until much soil has been eroded.

Gully erosion is when water converges into small rivulets and streams and takes with it large amounts of soil.

Desertification results from soil degradation and the subsequent absence of vegetation in arid and semiarid areas. Desertification is due to both climate change and human activities.
There are two major causes of erosion other deforestation: overcultivation and overgrazing. **Overcultivation** occurs when many farmers till and plow their farms to control weeds and increase crop yields. This practice exposes the soil to the wind and rain, which causes erosion. There are many techniques farmers can use to lessen the effect of overcultivation. These techniques include

- no-till or low-till farming, which is when the farmer uses special equipment to turn over the soil while covering it back up and planting new crops at the same time. Waste crops also are used as mulch during the next planting.

11. What is one way to reverse the problem of soil salinization?

A. add acid to lower the soil pH
B. add pesticides to increase crop yields
C. change to crops that can survive large amounts of salt
D. add large amounts of water to leach out the salts
E. add fertilizers to increase soil fertility

12. Farmers would choose which soil type for their crops?

A. sand
B. silt
C. clay
D. loam
E. mud

Free Response Questions

1. Describe the problems of soil erosion and desertification.

2. List and explain three ways a person could control erosion.

3. You pick up a sample of soil and decide you want to find out what kind of texture it has. Using a soil-texture kit, you determine that the soil is 20% sand, 60% silt, and 20% clay. Use the soil texture triangle to determine the texture of your soil.

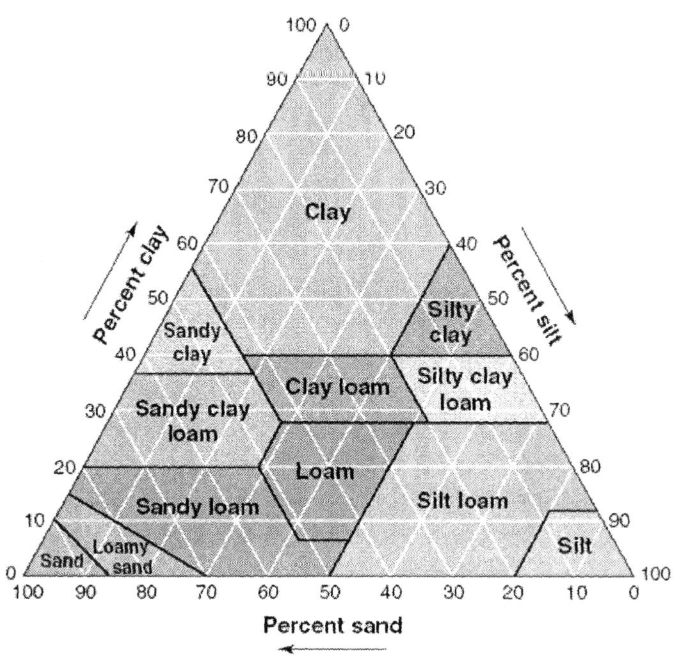

Multiple Choice Answers and Explanations

1. E. Inorganic fertilizers do not help with water-holding capacity.

2. C. Deforestation increases soil erosion.

3. C. Adding organic fertilizers such as compost to the soil lessens soil erosion. Organic fertilizers add to the soil and, by covering it, protect it from wind and rain.

4. B. Soils in tropical rain forests are not very deep and, because all of the nutrients are in the plants, are considered poor soils. Therefore, when a rain forest is clear cut, the nutrients are quickly depleted.

5. A. Surface litter is known as the O horizon.

6. D. The B horizon, also known as subsoil, has many minerals such as iron, aluminum, and calcium that are leached from the layers above.

7. A. light-colored soil usually contains a lot of sand and is low in nutrients.

8. D. Overgrazing is one of the main contributors to desertification.

9. A. Organic fertilizers have good water-holding capacity.

10. A. Mollisols are fertile, dark soils found in temperate grassland biomes and are considered the best agricultural soils.

11. D. Adding large amounts of water to crops will flush the salt out of the soil; however, this technique has many negative ecological results.

12. D. Loam, which is 40% sand, 40% silt, and 20% clay, is considered a good soil type by farmers.

Free Response Answers

1. Soil erosion and desertification are problems associated with overgrazing and overcultivation. Erosion occurs when ranchers are allowed to graze their animals to the point at which the land cannot recover. The same problem happens when farmers use unsustainable farming practices that till up the land and leave it exposed to the elements. Salinization can lead to desertification when salt blows onto the land and causes plants to no longer grow. The loss of plant life leaves the soil with no cover to protect it against wind and rain.

2. Any of the farming practices below could be used to answer this question.

- No-till or low-till farming is when farmers use special equipment to turn over the soil, cover it back up, and plant new crops, all at the same time. This practice also uses waste crops as mulch during the next planting.

- Alley cropping is the practice of planting crops between strips of trees. The trees shelter the crops from the wind, while the roots of the trees help hold the soil in place.

- Shelterbelts, or belts of trees planted around crops, help shelter crops from wind.

- Contour farming is the practice of planting crops across slopes or, in the case of terracing, in "stair" formations to slow water erosion.

- Strip cropping is the practice of planting alternating rows of different crop types, for example, a row of tall crops next to a row of shorter crops or a row of crops with fibrous roots next to a row of crops with taproots.

3. According to the soil texture triangle, a soil that is 20% sand, 60% silt, and 20% clay is silt loam.

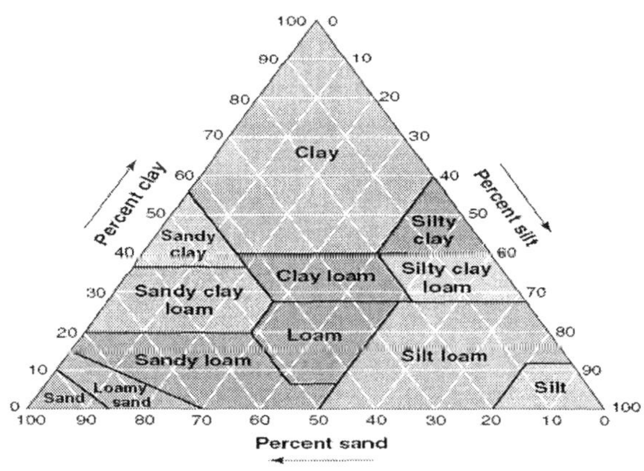

The Production and Distribution of Food

9.1 Crops and Animals: Major Patterns of Food Production

Feeding an ever-growing population has caused many problems, including

- the use of enormous quantities of fertilizers to maintain crop yields. Much of this fertilizer is washed away into groundwater and surface water, causing algal blooms and other forms of pollution.

- the increased use of pesticides and, consequently, the emergence of pesticide-resistant pests.

- the use of 70% of all the world's freshwater for unsustainable irrigation techniques.

However, it also has given rise to the **Green Revolution.** The increase in crop production, which marks the Green Revolution, has yielded the following benefits:

- the slowing of deforestation;

- the decrease in malnutrition, hunger, and infant mortality in many countries.

Subsistence farmers live on small parcels of land that provide them with just enough food for their households and maybe a little extra to sell for cash. Subsistence farmers seldom use inorganic fertilizers, relying instead on manure from their animals, which is enough to meet the nutrient needs of their family. This kind of agriculture works best in areas with a small population.

However, in many countries, the pressures of a growing population lead to a series of ecological consequences.

- Forests are clear-cut so as to create more land for crop production.

- Without forests, animal manure must be burned for fuel, rather than used as a fertilizer.

- The people must move farther away from their homes to gather food or firewood, leaving much of the soil susceptible to erosion.

- Good land is not given time to recover between plantings, but is instead planted over and over again, depriving the land of nutrients.

- All of these ecological changes increase poverty and put increased pressure on the land for food and income.

Many people, especially in the developed world, are turning to meat and dairy products for their food. The increase in animal farming has led to the following problems:

- the need to grow and use crops to feed livestock; one-fourth of the world's croplands and 70% of all grain crops in the United States are used to feed domestic animals.

- the overgrazing of much of the world's farmlands, making them susceptible to erosion and desertification.

- the contamination of surface waters with pathogens and algae blooms and the death of fish from the leakage of animal waste into streams and rivers.

- the frequent spread of diseases such as the recent avian flu among many industrial-style factory farms.

- deforestation for ranching and with it the release of additional carbon dioxide into the atmosphere, contributing to the greenhouse effect.

- an increase in the amount of methane (a greenhouse gas) in the atmosphere due to cow flatulence and belching. This methane makes up about 3% of the gases causing global warming.

9.2 From Green Revolution to Gene Revolution

Plant biotechnology refers to the process of genetically engineering different plants so that they exhibit certain traits. The goals of plant biotechnology are to

- help plants become resistant to certain diseases;
- help plants become resistant to environmental conditions such as drought or salinization;
- improve the nutritional value of crops;
- produce pharmaceutical products from ordinary plants;
- reduce the need for pesticides;
- reduce erosion;
- produce more food for a growing population.

While these goals are lofty, there are several concerns associated with the genetic engineering of plants, including

- environmental concerns, such as the concern that toxins will get into natural areas;
- food safety concerns, such as the concern that bioengineered organisms will cause an allergic response among some humans or that antibiotics will not work effectively on someone that has consumed genetically altered food;
- access concerns, such as the concern that farmers in developing countries will be unable to afford the higher costs of the genetically modified seeds.

9.3 Food Distribution and Trade

Most developed countries produce more food than they need for their population. Many developing countries are able to produce special crops (e.g., coffee, fruit, sugar, spices, cocoa, and nuts) that the developed world enjoys. The developed world exports their extra foodstuffs and imports the specialty items they are unable to grow or are out of season.

Food security, according to the Bread for the World's Hunger Institute, is the "assured access for every person to enough nutritious food to sustain an active and healthy life.")

9.4 Hunger, Malnutrition, and Famine

Hunger is a lack of the basic food required for energy and for meeting nutritional needs. Most hunger results from poverty.

Malnutrition is a lack of essential nutrients, such as specific amino acids, vitamins, and minerals.

Undernourishment is a lack of adequate food energy.

Famine is a severe shortage of food accompanied by a significant increase in the death rate. Causes of famine are

- drought
- conflict (war)
- Hunger Hot Spots:
    - Sub-Saharan Africa
    - Zimbabwe
    - Ethiopia
    - Sudan

Food Aid: donations mostly from US, Canada and Australia help but the problems are that free food undercuts local markets.

Multiple Choice Questions

1. What is/are reason(s) for hunger in the world today?

A. allowing good land to be left fallow and recover between plantings
B. drought and war
C. the diminishing of the sun's intensity in the last 100 years
D. the decrease in agricultural production around the world in the past 100 years
E. the Green Revolution

2. Most of the grain grown in the United States goes to

A. poor countries that cannot produce enough food for themselves
B. exports
C. aid organizations that distribute it to the poor
D. feed domestic animals
E. produce biodiesel

3. The benefits of bioengineered crops include all of the following *except*

A. resistance to plant diseases
B. increased genetic variety
C. reduced need for pesticides
D. reduced need for water
E. resistance to salinization

4. Why has the amount of fertilizers and the frequency with which they are used increased?

A. to increase crop yields
B. to conserve water
C. to eliminate erosion
D. to increase precipitation
E. to reduce ozone depletion

5. What is an effect of increased fertilizer use?

A. photosynthesis
B. eutrophication
C. overgrazing
D. salinization
E. malnutrition

6. People in developing countries must do many things to support themselves even as their populations continue to grow exponentially. Which of the following do these things include:

i. farm farther away from their homes
ii. burn manure for fuel due to deforestation
iii. continuously replant their crops season after season and not allow the soil to recover

A. i only
B. i and ii
C. ii and iii
D. i, ii, and iii
E. i and iii

7. Much of the increase in atmospheric methane is due to

A. photosynthesis
B. bovine flatulence
C. burning firewood for fuel
D. eutrophication
E. none of the above

8. Agricultural trade between developed and developing countries often involves which of the following:

i. Developing countries export staples.
ii. Developing countries export specialty crops.
iii. Developing countries import staples.
iv. Developing countries import specialty crops.
v. Developed countries export staples.
vi. Developed countries export specialty crops.
vii. Developed countries import staples.
viii Developed countries import specialty crops.

A. i, iv, v, and vii
B. ii, iii, vi, and viii
C. ii, iii, v, and viii
D. ii, iii, v, and vii
E. i, iii, vi, and vii

Free Response Questions

1. There has been a big shift in developing countries from sustainable agriculture to agriculture that is unsustainable.

a. Why are farmers in developing countries utilizing unsustainable farming practices?

b. What are the environmental impacts of such practices?
c. How might policy makers encourage sustainable farming practices?

Multiple Choice Answers and Explanations

1. B. Famine is a severe shortage of food accompanied by a significant increase in the death rate. Causes of famine are drought and conflict (war).

2. D. Seventy percent of grain grown in the United States is used to feed domestic animals.

3. B. Bioengineered crops do not increase genetic diversity, but they do

- help plants become resistant to certain diseases;
- help plants become resistant to environmental conditions such as drought or salinization;
- improve the nutritional value of crops;
- produce pharmaceutical products from ordinary plants;
- reduce the need for pesticides;
- reduce erosion;
- produce more food for a growing population.

4. A. Farmers have increased the amount of fertilizers and the frequency with which they use them to produce more crops and keep up with the food demands of a growing population.

5. B. Eutrophication occurs when plants are fertilized on the land. The excess fertilizer runs off into rivers, lakes, or streams, which causes an excess of algae. This algae, in turn, blocks out the sun, resulting in a lack of oxygen and the death of aquatic organisms.

6. D. People in developing countries must do many things to support their families, including moving farther away from their homes to gather food or firewood, burning manure for fuel, and planting crops one season after another to maximize growing time.

7. B. Methane is a by-product of the digestive process of ruminant animals. Cows and other animals eliminate 100 million tons of methane, a greenhouse gas, through belches and flatulence.

8. C. Developing countries export specialty crops and import staples, while developed countries export staples and import specialty crops.

Free Response Answers

1a. Farmers in developing countries are trying to feed their families and to make a little profit. In the short run, unsustainable agricultural practices are cheap and easy. For example, many farmers in Mexico will burn their crops after harvest so that they can begin to plant the next set of crops. A sustainable technique is to till the crop residue into the soil as fertilizer, rather than stripping the land of these needed nutrients.

1b. Some environmental impacts of such practices are deforestation, erosion, nutrient depleted soils, and sediment pollution in the water and air from eroded soil.

1c. Policy makers could encourage sustainable agriculture by providing subsidies or tax-incentives to farmers who use environmentally sound practices; by raising public awareness of the benefits of sustainable agriculture through educational outreach efforts; or by regulating the use of fertilizers and pesticides.

Wild Species and Biodiversity

10.1 The Value of Wild Species

Biodiversity refers to the "richness" of living species or the diversity of living things found in the natural world. It includes different species diversity, genetic diversity, and ecosystem diversity.

Species have two kinds of value:

- Instrumental value (benefits something else, usually humans)
    - Agriculture
    - Forestry
    - Aquaculture
    - Animal husbandry
    - Medicine/pharmaceuticals
    - Recreational
    - Aesthetic
    - Scientific
- Intrinsic value (value for it own sake)
    - Aesthetic
    - Scientific

10.2 Saving Wild Species

The **Lacey Act** (1900) forbids interstate commerce of illegally killed wildlife. Since the passing of this act, numerous refuges have been established. Today, the modifications to the act have been to prevent the importation of dangerous non-native species.

The **Endangered Species Act (ESA)** of 1973 (reauthorized in 1988) makes it illegal to kill, trap, uproot (plants), modify the habitat of, or engage in commerce of an endangered species or its parts.

For the Fish and Wildlife Service or the National Oceanic and Atmospheric Administration to designate a species as endangered or threatened, they must

1. list the species,
2. designate the critical habitat areas where the species is found,
3. develop a recovery plan to help the species survive and thrive.

Endangered species share the same characteristics:

- low reproductive rate;
- specialized feeding habits;
- feed at high trophic levels;
- large size;
- specialized nesting or breeding areas;
- found only in one place or region;
- have fixed migratory patterns;
- prey on livestock or people;
- have particular behavioral patterns;
- rare;
- commercially valuable.

10.3 Biodiversity and Its Decline

The following table identifies five of the primary contributors to the loss of biodiversity.

Habitat destruction	The greatest reason for the decline in biodiversity is the loss of habitat due to humans by • conversion: converting natural areas to farms, houses, etc. • fragmentation: breaking up one habitat with humans living in between • simplification: planting monocultures, removing dead branches, etc., from an environment.
Human overpopulation	The expansion of the human population, especially in the developing world, continues to harm natural ecosystems by encroaching on their habitats, overgrazing, etc.
Pollution	Pollution caused by human activity (e.g., oil spills, human wastes, acid deposition, pesticide use) contributes to the loss of biodiversity.
Exotic species	Exotic species, or species introduced into an area from somewhere else, can become invasive by outcompeting native species, having no natural predators, taking over an environment, and, often, destroying existing animals or plants.
Overuse	Over-hunting, illegal trade, and the exotic pet industry have contributed to a depletion of natural ecosystems by depleting the numbers of wild animals and plants.

10.4 Protecting Biodiversity

CITES (1970's), or the *Convention on Trade in Endangered Species of Wild Fauna and Flora*, focused on trade in wildlife and wildlife parts. In 1990, international trade on ivory was banned.

CBD (1992), or the Convention on Biological Diversity, is an international agreement that calls on various actions and steps between nations be taken to protect the world's biodiversity.

Multiple Choice Questions

1. The best way to protect biodiversity is to

A. pass endangered species laws.
B. protect world's ecosystems.
C. build more zoos.
D. import species from other countries.
E. use captive breeding and re-release programs.

2. The greatest threat to most species is

A. destruction of habitats.
B. air pollution.
C. invasive species.
D. build-up of toxic substances in food chains.
E. poaching.

3. Why do nonnative invasive species often become pests?

A. They don't have many offspring.
B. They increase biodiversity.
C. They can outcompete native species.
D. They stay in one area and take it over.
E. They can have many predators that eat them as prey.

4. A species is more likely to become endangered if it has which of the following?

i. Low biotic potential
ii. Small size
iii. Narrow distribution

A. i only
B. i and ii
C. iii only
D. i and iii
E. i, ii, and iii

5. The tiger is extinction prone for all of the following reasons *except*

A. low reproductive rate.
B. feeding at the top trophic level.
C. specialized feeding habits.
D. overfishing of the Ganges.
E. large size.

6. The organization that identifies and lists endangered species is the

A. Fish and Wildlife Service.
B. Environmental Protection Agency.
C. National Oceanic and Atmospheric Administration.
D. National Aeronautics and Space Administration
E. A & C

7. Building homes in the middle of a wildlife preserve might lead to which of the following?

A. an increase in wildlife reproduction rates
B. a reduction in nonnative invasive species
C. an increase in the population of top predators
D. a reduction in biodiversity
E. an increase in biological corridors

8. In contrast to the Endangered Species Act, the Convention on Biological Diversity

A. seeks to limit the decline in biodiversity through political means.
B. provides protection for endangered species' habitats.
C. is international in scope.
D. is subject to political resistance.
E. is without opposition.

9. In 2000, 20,000 seals in the Caspian Sea died from canine distemper. What was this death attributed to?

A. global warming
B. simplification
C. pathogen pollution
D. overuse
E. hunting

10. All of the following are reasons for international protection of species *except*

A. trade in endangered species.
B. habitats often overlap international boundaries.
C. pollution in one country may affect species in another country.
D. poor countries often do not have the economic resources to manage species' habitats.
E. it is important that the United States lead the world in environmental conservation.

Free Response Questions

1. Identify two processes that cause physical alteration of habitat and explain how they affect biodiversity.

2. Read the following statement from U.S. Representative James Hansen of Utah:

> "The ESA has become a wrecking ball in this country, devastating personal finances and regional economies."

What factors contribute to the support of or opposition to environmental legislation?

Multiple Choice Answers and Explanations

1. B. If we protect the world's ecosystems, none of the other measures would need to be enacted.

2. A. The greatest reason for the decline in biodiversity is the loss of habitat.

3. C. Nonnative invasive species can become invasive by outcompeting native species, having no natural predators, taking over an environment, and, often, destroying native animals or plants.

4. D. The characteristics of endangered species are low reproductive rate (i.e., low biotic potential), large size (not small size), found only in one place or region, specialized feeding habits, feed at high trophic levels, specialized nesting or breeding areas, fixed migratory patterns, prey on livestock or people, have particular behavioral patterns, and are rare or commercially valuable.

5. D. Tigers are extinction prone for all the reasons except overfishing in the Ganges.

6. E. Both the Fish and Wildlife Service and the National Oceanic and Atmospheric Administration identify and list endangered species.

7. D. Building homes in the middle of wildlife preserves can fragment the area. This fragmentation of the land can lead to a decrease in biodiversity because the animals can no longer interbreed, migrate, etc.

8. C. The Endangered Species Act is specific to the United States, while the Convention on Biological Diversity is international in scope.

9. C. The seals died from an infection spread by local dogs.

10. E. The need for international protection of species is necessary because of trade in endangered species, because habitats often overlap international boundaries, because pollution in one country may affect species in another country, and because poor countries often do not have the economic resources to manage species' habitats. It is not necessary for the United States to lead the world in environmental conservation.

Free Response Answers

1. There are three processes that cause physical alteration of habitat and can affect biodiversity: conversion, fragmentation, and simplification. You need only explain two of these. Conversion refers to converting natural areas to farms or houses. Any loss of natural habitat can result in only one thing: a proportional reduction in all populations that require that habitat. Fragmentation refers to breaking up a habitat into two or more habitats separated by human settlements. Fragmentation can prevent migration between the fragments, a greater proportion of edges (which can be detrimental to some species), and a disruption in mating opportunities. Simplification refers to practices or activities that simplify a habitat, such as planting monocultures or removing dead branches or trees. When humans simplify the environment, they remove an important microclimate that many species need to survive. This removal can lead to a reduction in biodiversity.

2. Support for environmental legislation comes from environmental groups looking to sustain natural habitats and develop ecologically secure and long-lasting environments. Opposition comes from development, timber, and mineral interests looking to maximize profits and contain costs. These groups feel that the Endangered Species Act places more importance on the endangered species than on human needs and economic growth.

Ecosystem Capital: Use and Restoration

11.1 Global Perspective on Biological Systems

Natural resources are natural ecosystems and the biota in them. These resources have both economical and ecological value.

11.2 Conservation, Preservation, Restoration

A **renewable resource** is any resource that can be renewed by reproduction and regrowth, such as trees.

Conservation	To regulate the resource so that it is not depleted For example, if we conserve trees, we cut them down at a rate that allows them to regrow and recover.
Preservation	To preserve the resource in the natural state it is in For example, if we preserve a forest, we never cut it down.

Consumptive Use	When people harvest natural resources for their own personal use, for example, for food, for shelter, or for clothing
Productive use	To use a resource for economic gain, for example, cutting down trees to sell for wood

Restoration (reclamation) is repairing an ecosystem after it has been harmed. The difficulty of restoration depends on the extent and type of damage that has occurred.

The **maximum sustainable yield** (MSY) is the highest rate of use a resource can handle while replacing or sustaining itself.

Carrying capacity is the maximum population an ecosystem can support on a sustainable basis. (See also Chapter 4)

Fig. 11.1

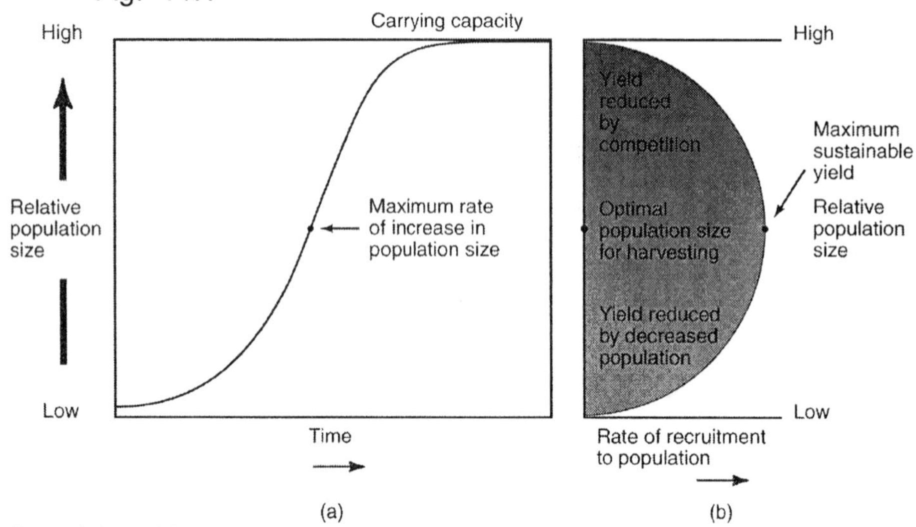

Figure 11-6 Environmental Science, 10/e
© 2008 Pearson Prentice Hall, Inc.

The phenomenon known as the **tragedy of the commons** occurs when a resource utilized by many people is exploited and eventually depleted. For example, many fishing companies take all the fish they can each season, which leads to a general depletion of fish stocks. Because the companies do not own the ocean and because they are trying to maximize their profits, they do not limit how many fish they catch. If one company does not take as many fish as it can, then some other company will take more and, in doing so, gain an economic advantage. Garrett Hardin described this phenomenon in his essay *The Tragedy of the Commons.*

There are two solutions to the tragedy:

- private ownership. The assumption is that if you own the resource, you will take better care of it.

- regulation of the commons. Establish laws that prevent tragedy from occurring, for example, limits on how many fish can be caught per company per season.

11.3 Biomes and Ecosystems under Pressure

Forest Biomes

Thirty percent of the land area of the world is forest. Forests provide the following ecological services:

- conservation of biodiversity;
- moderation of climate;
- prevention of erosion;
- storage of carbon and nutrients;
- replenishing oxygen

Deforestation refers to removing a forest so that the land can be used for other purposes. Logging is not considered deforestation if the trees are replanted. The following are consequences of deforestation:

- reduction in productivity of the forest;
- reduction in nutrients available;
- reduction in biodiversity;
- increases in erosion;
- increases in runoff;
- reduction in the amount of CO_2 that is removed from the atmosphere;
- reduction in forest products;
- loss of livelihood for people depending on a forest for survival.

One study found that the greatest cause of deforestation was due to the conversion of forests into pastures or agricultural fields.

There are four forest management methods:

- **Clear-cutting** is removing all trees in an area at one time. This method lowers biodiversity and leaves a barren landscape.

- **Selective cutting** is removing some trees. This method leaves a forest with high biodiversity and beauty.

- **Shelter-wood cutting** is removing mature trees in groups over many years and leaving many trees to reseed the area.

- **Seed-tree cutting** is removing all the small trees and leaving the large, mature, seed-bearing trees.

Sustainable forest management means that forests are managed as ecosystems and that maintaining the biodiversity and function of the ecosystem is the primary objective.

Ocean Ecosystems

Many marine fisheries are being overfished. According to the U.N.'s Food and Agricultural Organization (FAO),

- 52% of fish stocks are fully exploited,
- 16% are overexploited,
- 8% are depleted.

Aquaculture is the practice of growing fish and other aquatic organisms in large tanks. The hope is that this fishing method will help to eliminate poverty and improve the lives of people by allowing them to produce valuable food. This method, however, does cause some environmental problems such as pollution of rivers.

Coral reefs are one of the most diverse ecosystems in the world. Corals are animals that have a endosymbiotic algae living in their tissues (zooxanthellae) that allows them to photosynthesize. Corals live in waters that are clear, warm, and shallow enough for photosynthesis. When corals lose their algae (usually due to warm waters from El Niño or global warming), they turn white, a problem called coral bleaching. Corals are in danger, especially from the exotic pet trade. In an attempt to capture and sell coral, people dynamite the reefs or use cyanide to stun fish. These activities destroy reefs.

Mangroves are salt-tolerant trees found on the coasts of many countries. These trees protect the coast from erosion and are a sanctuary and nursery for many fish and invertebrate species. Mangroves are being cut down at an alarming rate, primarily for wood and to make room for coastal land development.

11.4 Public and Private Lands in the United States

The United States has set aside 40% of its land for public use. Wilderness areas are given the greatest amount of protection (4% of the United States). Permanent structures, roads, and cars are not allowed on this land.

Old-growth forests are forests composed of old trees (hundreds of years). Only 5% of forests in the United States are considered old-growth forests.

Second growth forests are the most common forest found in the United States.

Multiple Choice Questions

1. Litter along the highway is an example of

A. consumptive use.
B. productive use.
C. restoration.
D. tragedy of the commons.
E. private ownership.

2. Cutting all small trees but leaving a few mature seed-producing trees is referred to as

A. selective cutting.
B. clear-cutting.
C. shelterwood cutting.
D. seed-tree cutting.
E. deforestation.

3. A subsistence farmer in a developing country who is trying to maximize the amount of crops that he can grow, while disregarding biodiversity, would most likely use what method?

A. selective cutting
B. clear-cutting
C. shelterwood cutting
D. seed-tree cutting
E. restoration

4. Establishing a wildlife refuge would be described best as

A. conservation.
B. preservation.
C. restoration.
D. consumptive use.
E. productive use.

5. Which of the following would most likely result in the extinction of species?

A. loss of habitat
B. invasive species
C. pollution
D. hunting
E. global climate change

6. A key factor in the decline of world fish stocks is

A. coral bleaching.
B. overfishing.
C. competition from aquaculture.
D. water pollution.
E. surface runoff.

7. All of the following illustrate the tragedy of the commons *except*

A. overgrazing of public land.
B. air pollution.
C. depletion of ocean fisheries.
D. depletion of an aquifer by local farmers.
E. dumping on private land.

8. Which of the following would be an example of reclamation of disturbed lands?

A. converting an abandoned strip mine into a theme park
B. converting a forest into pastureland
C. making it illegal to hunt in national parks
D. flooding the everglades by removing manufactured dams
E. leaving a forest undisturbed and natural

9. Which of the following would most likely occur in an area that had been clear-cut?

A. Soil erosion would decrease.
B. The climate in the area would stay the same.
C. Water runoff would increase.
D. The depth of the soil would increase.
E. The nutrients in the soil would increase.

10. One reason coral reefs are declining is

A. the exotic pet industry.
B. colder ocean waters.
C. decreased sunlight.
D. continental drift.
E. radon gas emissions.

Free Response Question

1. Compare and contrast the advantages and disadvantages of clear-cutting a forest with the use of selective cutting.

Multiple Choice Answers and Explanations

1. D. The tragedy of the commons occurs when a resource not owned by an individual but utilized by many people is exploited and eventually depleted.

2. D. Seed-tree cutting is removing all small trees while leaving large, mature, seed-bearing trees in a forest.

3. B. Clear-cutting is removing all trees in an area at one time. This method of forest management lowers biodiversity and leaves a barren landscape.

4. B. Establishing a wildlife refuge would involve leaving a resource in its natural state; thus, this is an example of preservation.

5. A. Loss of habitat is the single greatest cause of the loss of biodiversity.

6. B. Overfishing has led to a huge decline in fish stocks in the world's oceans.

7. E. Tragedy of the commons only refers to public land.

8. D. Reclamation is returning something back to its natural ecosystem after it has been destroyed. Damning a river dries up the everglades, so tearing the dams down would be reclaiming the everglades to their natural state.

9. C. Water runoff would increase in an area that has been clear-cut because the trees would not be there to absorb the water.

10. A. The pet industry is a large contributor to coral reef destruction. Many people dynamite or use cyanide to capture fish around coral and, in doing so, destroy the reef.

Free Response Answers

1. Although clear-cutting is cheaper and more efficient, it destroys habitat and reduces biodiversity. Selective cutting is more costly than clear-cutting because large equipment like bulldozers cannot be used to cut down trees. However, selective cutting preserves the natural setting of the landscape, is more aesthetic, reduces erosion and pollution problems, and maintains biodiversity.

Energy from Fossil Fuels

12.1 Energy Sources and Uses

Throughout history, energy has come from many sources:

- human or animal power;
- coal (during the Industrial Revolution to power steam engines);
- oil (cleaner burning than coal and has more advantages in engines);
- natural gas.

The three fossils fuels—oil, coal, and natural gas—provide 85% of the United States' energy and 87% of the world's energy.

<u>Electrical Power</u>

The most common way to get electricity is to use a **turbogenerator** (Figure 12–7a). A turbogenerator works by

- Creating a high-pressure steam by boiling water,
- Directing the high-pressure steam at a turbine (like a big fan blade) to cause the blades to turn,
- Connecting the turbine to a generator, which turns mechanical energy into electrical energy.

Any primary energy source (coal, oil, or natural gas) can be used to boil the water. (A renewable energy source or nuclear power can be used to turn the turbine. These renewable energy sources will be discussed in Chapters 13 and 14.)

Figure 12-7a

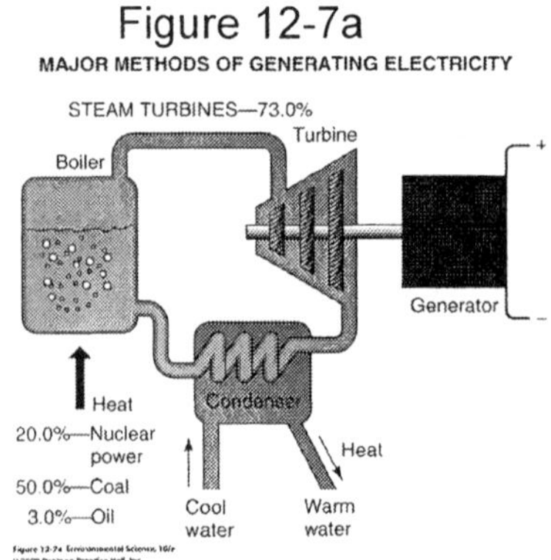

Producing electricity is only about 30% efficient; 70% of the energy is lost to heat (second law of thermodynamics). The heat generated is usually put into the environment through the air or water. The latter, called thermal pollution, can harm aquatic organisms.

12.2 Exploiting Crude Oil

As living organisms die, they get buried under layers of sediment. Heat and pressure convert the remains of living organisms into coal, oil, and natural gas. This process is slow; it takes millions of years to produce the amount of energy the world consumes in one day around.

Estimated reserves are educated guesses about how much oil or natural gas is left and where it may be found.

Proven reserves are exact amounts of oil that can be economically obtained from a field. Proven reserves are determined only after drilling.

There are three methods for recovering oil.

- Primary recovery through conventional pumping removes only 25% of the oil.

- Secondary recovery, which involves injecting brine or steam into a well to pressurize and force out the remaining oil, removes 50% more of the oil from the field.

- With enhanced recovery, carbon dioxide is injected into the oil field, which breaks up oil droplets and allows more oil to be removed.

The United States has become increasingly dependent upon foreign oil reserves. There are several problems associated with this dependence.

- Thirty percent of the United States' balance-of-payment deficit goes to oil. (The price of a barrel of oil has tripled in the last six years.)

- The Middle East (where most of our nation's oil is imported) is politically unstable, and a strong U.S. military presence will have to be maintained to ensure access to this oil.

- Domestic supplies of oil are running low, and no new oil fields are expected to be found.

There are ways the United States can extend its oil supplies and reduce its dependence on foreign oil.

- It can increase the fuel efficiency of vehicles.
- It can utilize fossil fuels other than just oil.
- It can develop alternative fuels.

12.3 Other Fossil Fuels

Natural Gas

- The United States has a 25-year supply of natural gas with an estimated reserve of 50 years.
- Fifteen percent of the natural gas Americans use is imported, mainly from Canada.
- Cars can be run off natural gas, which is cleaner to burn than oil.

Coal

- In the United States, 50% of electricity comes from coal-fired power plants.
- There is approximately 225 years of coal left in proven reserves.
- Strip mining, the most common method of coal removal, is very degrading to the natural ecosystem. While the land can be reclaimed and brought back, reclamation takes years.
- Coal goes through a series of stages as heat and pressure increases and moisture content decreases. Those stages are, in order, peat, lignite, bituminous, and anthracite.

Oil Shales and Sands

- **Oil shale** is a rock that, when heated to 1,100°F, produces a type of crude oil that can be extracted and refined into gasoline. The organic matter contained within the oil shale is called kerogen.
- **Oil sands** are rocks containing a tar-like substance that, when heated, can be refined.
- Producing oil from oil shale and oil sands is expensive and damaging to the environment.

Cogeneration refers to using a single energy source to produce both electrical and heat energy. This is known as CHP (combined heat and power).

12.4 Energy Security and Policy

Currently, there are three main security threats that shape the United States' energy use.

- oil dependence. Sixty-five percent of proven oil reserves are in the Persian Gulf region.

- terrorism. Nuclear power plants, dams, oil and gas pipelines, refineries, tankers, and electrical grids are targets for terrorism.

- global climate change. If we continue to produce greenhouse gasses (especially carbon dioxide), polar ice caps will melt, which will cause sea levels to rise, changes in rainfall, more extreme storms, droughts, heat waves, and ecosystems to change.

Multiple Choice Questions

1. A power company uses natural gas to run a plant and then uses the waste heat to produce electricity. This is an example of

A. oil dependencc.
B. primary recovery.
C. cogeneration.
D. reclamation.
E. turbogenerators.

2. If the average American uses 23 barrels of petroleum a year, then how much does the entire population use?

A. 10 million barrels
B. 250 million barrels
C. 2 billion barrels
D. 10 billion barrels
E. 7 billion barrels

3. Which statement about coal supplies is true?

A. South America and Central America contain 50% of the world's coal supplies.
B. North America contains 40% of the world's coal supplies.
C. The Middle East contains 33% of the world's coal supplies.
D. Eastern Europe and the former U.S.S.R. contain 54% of the world's coal supplies.
E. North America, Eastern Europe, and the former U.S.S.R. contain 56% of the world's coal supplies.

4. The stages of coal, classified by heat and sulfur content, from the softest and most moist to the hardest and driest are

A. bituminous, anthracite, peat, lignite.
B. lignite, anthracite, peat, bituminous.
C. bituminous, peat, lignite, anthracite.
D. peat, lignite, bituminous, anthracite.
E. peat, bituminous, anthracite, lignite.

5. Which of the following countries has the largest coal reserves?

A. Japan
B. Italy
C. China
D. Mexico
E. Africa

6. Approximately how many years of proven reserves of coal are left?

A. 50 years
B. 75 years
C. 125 years
D. 175 years
E. 225 years

7. Which of the following is an example of cogeneration?

A. using solar panels to convert light energy into heat
B. using wave action to generate electricity
C. using a dam to convert hydraulic energy into electrical energy
D. using heat from a waste incinerator to generate electrical energy
E. using radioactive fission to produce electricity

8. Which energy source is the cleanest in the United States?

A. oil
B. lignite
C. natural gas
D. radioactive material
E. anthracite

9. Which of the following is not a reason for importing oil from the Middle East?

A. It is cheap.
B. It is plentiful.
C. It is easily accessed.
D. It is higher quality than other crude oils.
E. It is the least environmentally sensitive region for oil production.

10. Strip mining is detrimental to the environment for all of the following reasons *except*

A. contamination of groundwater.
B. disruption of habitat.
C. air pollution.
D. aesthetic appeal.
E. destruction of wetlands.

Free Response Questions

1. Running at full capacity, a power company is able to generate 10 megawatts of electricity to a city that demands 10,000 kilowatt hours (kWh) of electrical energy each year. There are 5,000 homes in the city, and each homeowner pays $0.10 per kWh.

a. If the power plant operates at full capacity for 10,000 hours/year, how many kilowatts of electricity can be produced by the plant in a year?

b. At the current rate of electrical energy use per household, how many kilowatts of electrical energy does the community consume in one year?

c. Identify two environmental effects that coal-burning power plants are having on the environment.

Multiple Choice Answers and Explanations

1. C. Cogeneration refers to using a single source to generate both electrical energy and heat energy.

2. E. There are 300 million Americans. When you multiply the population by 23 barrels of petroleum per year, you get approximately 7 billion barrels of petroleum per year.

3. E. Twenty-eight percent of the world's coal supplies are located in North America, and 28% are located in Eastern Europe and the former U.S.S.R.

4. D. The stages of coal from softest to hardest and most moist to least moist are peat, lignite, bituminous, and anthracite.

5. C. China has the most coal of these countries.

6. E. There is approximately 225 years of known coal reserves.

7. D. Cogeneration refers to using a single source to generate both electrical energy and heat energy.

8. C. Natural gas is the cleanest burning fuel available in the United States.

9. E. The Middle East is a fragile marine and desert ecosystem.

10. E. Strip mining is the least destructive to wetland habitats because most mining is not on the coast of countries.

Free Response Answers

1a. 1.0×10^4 kW X 10,000 = 100,000,000 kWh/yr or 1.0×10^8

1b. 5,000 homes X 10,000 kWh = 50,000,000

1c. A coal-burning power plant contributes to many environmental hazards. These hazards include (but are not limited to) high emissions of carbon dioxide, sulfur dioxide, dust, particulates, thermal pollutants, and NO_x. In addition, coal mining pollutes the environment with acids and tears up the land.

Energy from Nuclear Power

13.1 Nuclear Energy in Perspective

- Sixteen percent of the world's energy comes from nuclear power.

- Thirty-one countries operate nuclear power plants.

13.2 How Nuclear Power Works

In **fission,** a large atom of one element is split to make two smaller atoms of different elements (Fig. 13–5a-5b). In **fusion,** two smaller atoms combine to form a larger atom of a different element (Fig. 13–5b).

Figure 13-5a-5b

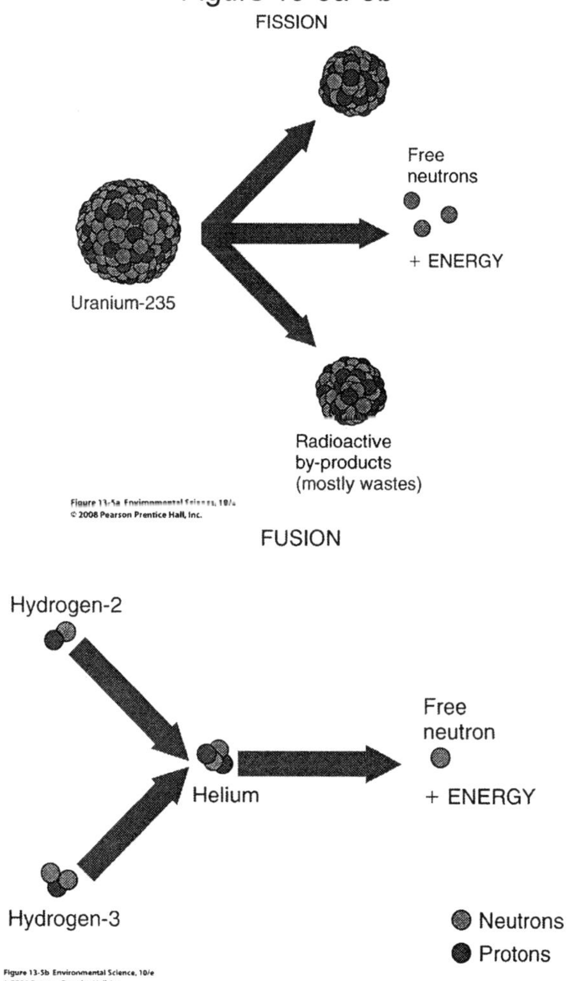

In a nuclear power plant, heat from a reactor is used to boil water to provide steam that drives a turbogenerator, as illustrated in Figure 13–8.

Figure 13-8

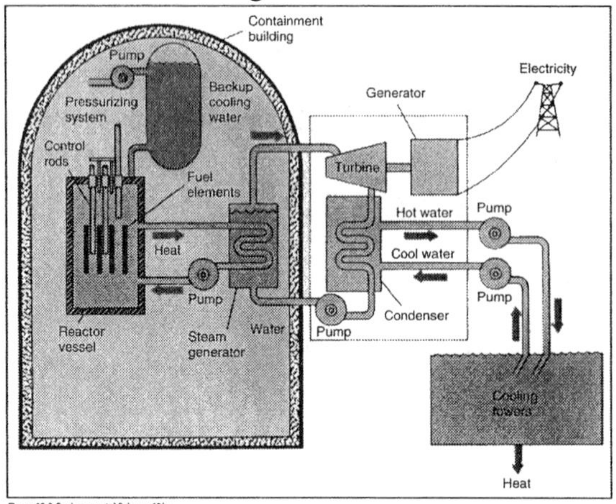

Figure 13-8 Environmental Science, 10/e
© 2008 Pearson Prentice Hall, Inc.

How does a nuclear power plant compare to a coal-fired power plant? The following chart compares the two in terms of fuel needed, carbon dioxide emissions, sulfur dioxide and other emissions, radioactivity, solid wastes, and accidents.

COAL	NUCLEAR
Two to three millions tons of coal is needed.	Thirty tons of enriched uranium is needed.
Seven million tons of carbon dioxide is emitted into the atmosphere.	No carbon dioxide is emitted into the atmosphere except to mine and to enrich uranium.
Three hundred thousand tons of sulfur dioxide and other pollutants that lead to acid rain is produced.	No acid-forming pollutants are produced.
One hundred times more radioactivity than a nuclear power plant is released because of radioactive compounds in the coal.	Only low-levels of radiation are released as waste gas.
Six hundred thousand tons of ash that must be buried is released.	Two hundred and fifty tons of highly radioactive wastes requiring safe storage is produced.
Fire or fatalities to workers can occur.	If an accident occurred, catastrophic releases could result that lead to widespread radiation sickness, human deaths, cancers, and environmental contamination.

13.3 The Hazards and Costs of Nuclear Power Facilities

A **high-dose exposure** to radiation can destroy cells and cause immediate death.

A **low-dose exposure** to radiation can lead to malignant tumors or leukemia, damage to eggs or sperm causing birth defects, a weakening of the immune system, mental retardation, and cataracts.

Radioactive Waste

Half-life is the time it takes for half of the amount of a radioactive isotope to decay. Each isotope has a half-life that ranges from a second to thousands of years.

Disposal of Radioactive Wastes

Short-term containment is used to store short-lived isotopes. In 10 years, fission wastes lose more than 97% of their radioactivity.

- Deep swimming pool-like tanks on the site of the nuclear power plant are where spent fuel is first stored.
- By 2015, U.S. nuclear storage sites will be full.
- After a few years, spent fuel can be placed in air-cooled dry casks for storage, but this option is expensive.

Long-term containment is used to store long-lived isotopes. The EPA suggests a minimum of 10,000 years of isolation to provide protection from long-lived isotopes.

- Most countries bury high-level radioactive waste deep underground.
- Beginning in 2018, the United States will begin burying its high-level radioactive waste in Yucca Mountain, which is located in southwestern Nevada.
- NIMBY refers to *not in my backyard.* Most people including those living in Nevada do not want nuclear waste stored in their neighborhood.

Nuclear Power Accidents

In 1979, the **Three Mile Island** nuclear power plant in Harrisburg, Pennsylvania, experienced a partial meltdown due to human and equipment failures and a flawed design. Ten million curies of radioactive gas were released into the atmosphere.

In 1986, at the **Chernobyl** power plant in Ukraine, engineers disabled the power plant's safety system to test the diesel generators. This action decreased the flow of coolant water in the reactor. The lack of coolant caused the reactor to heat up. Engineers quickly inserted carbon-tipped control rods to slow down the reaction. However, steam explosions blew the 2,000-ton top off the reactor, the reactor partially melted down, and a fire was started. Four hundred times more radiation than the bombs of Hiroshima and Nagasaki together was released into the atmosphere. Hundreds of thousands of people are expected to die from cancer as a result of being exposed to this radiation. Already, many children who drank contaminated milk have suffered from thyroid cancer.

Terrorism

Fears of terrorism loom over nuclear power. If a terrorist group decides to target a nuclear power plant, the results would be deadly. The Energy Policy Act of 2005 has stepped up security measures at nuclear power plants across the nation.

Economics

The building of all types of power plants have been postponed due primarily to the following economic reasons:

- Demand for electricity is less than expected.
- Safety standards for nuclear power plants make building costs very high.
- Public protests have delayed construction.
- If an accident were to occur, the clean up would cost billions of dollars.
- Nuclear power plants cannot stay open very long due to embrittlement and corrosion. The amount of money a plant can generate before it must be closed (on average about 17 years) is limited.

Multiple Choice Questions

1. Compared to a coal-fired power plant, a nuclear power plant generates more

A. thermal pollution.
B. soot.
C. carbon dioxide.
D. sulfur dioxide.
E. particulates.

2. A problem associated with nuclear power plants is

A. production of carbon dioxide.
B. production of sulfur dioxide.
C. disposal of radioactive waste.
D. contributions to global warming.
E. contributions to acid rain.

3. Uranium-235 has a half-life of 710 million years. If a certain amount is considered safe after the radioactivity has dropped to 0.10 percent of the original level, how long must the Uranium-235 be stored to be safe?

A. 7.1×10^5 years
B. 7.1×10^6 years
C. 7.1×10^7 years
D. 7.1×10^8 years
E. 7.1×10^9 years

4. Which of the following metals is considered an energy resource?

A. uranium
B. aluminum
C. iron
D. copper
E. mercury

5. A sample of radioactive waste has a half-life of 20 years and an activity level of 4 curies. After how many years will the activity level of this sample be 0.50 curies?

A. 20 years
B. 30 years
C. 40 years
D. 50 years
E. 60 years

6. In contrast to low-level radioactive waste, most high-level radioactive waste in the United States is currently

A. stored at Yucca Mountain
B. incinerated
C. stored at reactor sites
D. put into landfills
E. recycled

Free Response Questions

A nation conscience of both environmental and economic concerns is deciding whether to invest in coal-fired power plants or nuclear power plants.

1. What are the environmental costs associated with each?

2. What are the environmental benefits associated with each?

3. What are the economic costs associated with each?

4. What are the economic benefits associated with each?

Multiple Choice Answers and Explanations

1. A. Compared to a coal-fired power plant, a nuclear power plant generates an extreme amount of heat, which is usually put into a water source and causes thermal pollution.

2. C. Storage and disposal of nuclear radioactive waste is a huge concern of the nuclear power industry.

3. E. 7.1×10^9 because $\frac{710 \times 10^6}{1} = \frac{X}{0.1}$ (or 10 half lives)

4. A. Uranium is used to generate nuclear power along with other radioactive elements.

5. E. In 20 years, the radioactivity will be half or 2 curies. After 20 more years, it will be half again or 1 curie. And after another 20 years, it will be half again or .50 curies. Twenty years plus 20 years plus 20 years is 60 years.

6. C. Currently, most radioactive waste in the United States is stored at reactor sites until Yucca Mountain is complete or another solution is found.

Free Response Answers

1. There are environmental costs associated with both nuclear power plants and coal-fired power plants. Nuclear power plants produce 250 tons of highly radioactive waste that requires safe storage. If an accident were to occur and radioactive waste were to be released into the atmosphere, widespread radiation sickness, human deaths, cancers, and environmental contamination would result. Coal power plants, on the other hand, emit 7 million tons of carbon dioxide into the atmosphere, produce 300,000 tons of sulfur dioxide and other pollutants that lead to acid rain, and release 600,000 tons of ash that must be buried.

2. In comparison to coal-fired power plants, nuclear power plants are cleaner burning and produce little or no greenhouse gasses. Unlike with nuclear power, a nation using coal does not have to deal with either disposing nuclear waste, or with acts of terrorism that target their energy source.

3. Nuclear power plants are expensive to build and repair and have a shorter lifespan. In addition, an environmental clean up after a nuclear disaster would cost billions of dollars. The economic costs of coal are associated with environmental clean up and pollution prevention.

4. There are not many economic benefits associated with nuclear power, but the following may be considered: no pollution prevention devices like scrubbers are needed (though offset by containment issues), and the amount of radioactive substance needed may be cheaper to extract than coal. Coal, on the other hand, is relatively inexpensive and plentiful.

Renewable Energy

14.1 Putting Solar Energy to Work

Active solar systems use pumps, photovoltaic cells, and other technologies to generate electricity or heat. Solar cells are about 20% efficient.

Passive solar systems use natural systems like the Sun, shades on windows, or landscaping to heat or cool a building.

Solar energy is more expensive than conventional energy sources.

Since solar energy is only available during the day, batteries or back-up sources are needed at night or during the winter months.

Solar energy is a nonpolluting and free source of clean fuel that could help to reduce the world's dependence on fossil fuels.

Photovoltaic cells are the fastest-growing source of energy in the world.

14.2 Indirect Solar Energy

Dams, firewood, windmills, and sails are all examples of indirect solar energy because they all get their energy from the Sun.

Hydropower is generated by hydroelectric dams, where water under high-pressure flows through channels that drive turbogenerators. About 6.7% of electricity in the United States comes from dams (17% worldwide). There are both pros and cons associated with hydropower.

- Dams flood and destroy landscapes above the dam but control flooding below the dam.
- Fish habitats can be destroyed by dams.
- Dams limit the amount of freshwater that flows downstream, which can harm estuaries.
- Dams produce no air pollutants.

With **wind power,** the wind is used to turn large turbines and drive turbogenerators. Wind power is the second fastest growing energy source in the world (1.3% of global electricity). There are pros and cons associated with wind power.

- The land beneath the turbines can be used for farming.
- Wind produces no air pollutants.
- Windmills destroy the aesthetics of an ecosystem.
- Windmills have been linked to the killing of migratory birds.

Biomass energy is energy produced from burning wood or plant residue or from burning organic wastes. There are pros and cons associated with biomass energy.

- Biomass energy is a renewable energy source.
- Although burning produces air pollutants, it is less damaging than fossil fuel emissions.
- It is a great use of wastes like sugarcane crops or landfill.
- In developing countries, firewood is the only source of fuel for cooking. This leads to deforestation

14.3 Renewable Energy for Transportation

Biofuels are fuels processed from organic matter, such as plant and animal wastes. Examples of biofuels include

- ethyl alcohol, which is produced by fermenting and then distilling sugar or starches.
- biodiesel, which is a fuel produced from left over vegetable oils.

Hydrogen is not a fuel but an alternative energy carrier, like electricity.

- Hydrogen burns cleanly with only water vapor as a product.
- There is not much hydrogen *gas* left on Earth, so hydrogen would have to be generated, which would take a lot of energy.
- Storing hydrogen in a vehicle is difficult.

Fuel cells, which store hydrogen and use it to produce electricity, are an alternative to burning hydrogen. Fuel cells are being used in some buses.

14.4 Additional Renewable-Energy Options

Geothermal energy is when heat from deep within the earth is used to heat buildings or drive turbogenerators.

Tidal power can be generated by building a dam across the mouth of a bay and using the incoming tide to turn turbines and drive turbogenerators. Although tidal power is a renewable energy source, only about 30 locations in the world have tides large enough to use tidal power effectively. The dams also trap nutrient rich sediments and stop the migration of fish.

Ocean thermal-energy conversion uses the temperature difference between the warm surface ocean waters and the colder deep waters to generate power.

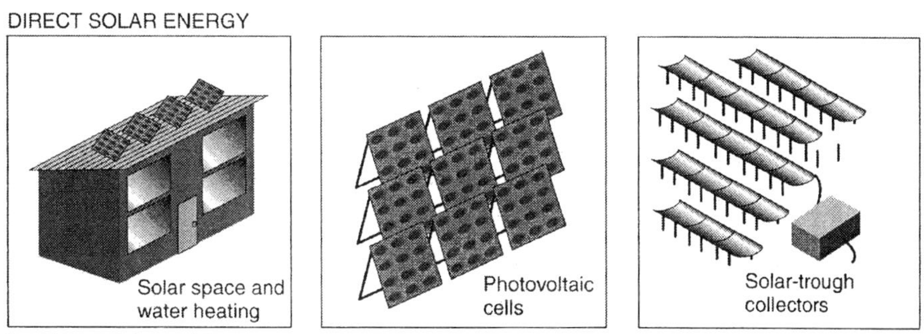

Figure 14-27 part 1 Environmental Science, 10/e
© 2008 Pearson Prentice Hall, Inc.

Figure 14-27 part 2 Environmental Science, 10/e
© 2008 Pearson Prentice Hall, Inc.

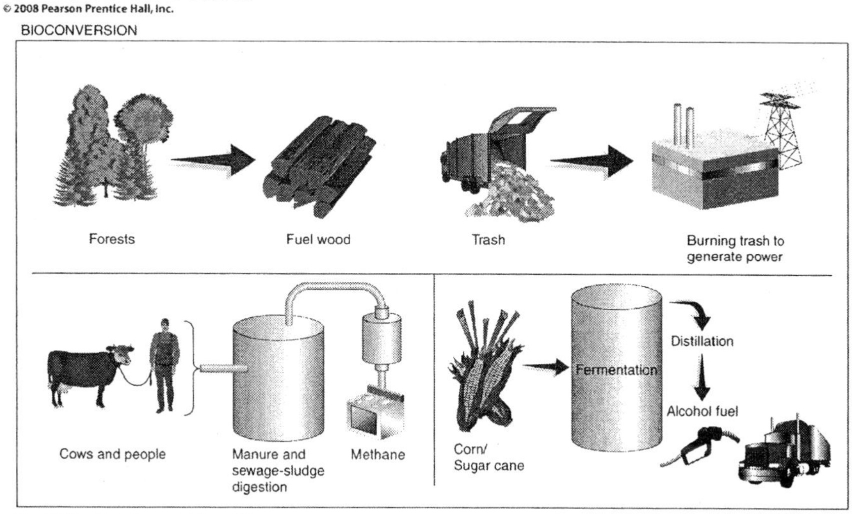

Figure 14-27 part 3 Environmental Science, 10/e
© 2008 Pearson Prentice Hall, Inc.

Multiple Choice Questions

1. Which of the following energy sources is not made directly or indirectly from the Sun?

A. wind
B. geothermal
C. solar
D. wind
E. biofuels

2. Hydroelectric power plants have many negative environmental impacts such as

A. flooding downstream.
B. air pollution.
C. protection of estuaries.
D. destruction of fish habitats and migration.
E. killing of migratory birds.

3. What supplies the most commercial energy in the world today?

A. solar power
B. geothermal energy
C. nuclear power
D. oil
E. hydroelectric energy

Match the following energy sources with their characteristics:

A. hydropower
B. wind power
C. biomass
D. geothermal
E. tidal

4. What energy source leads to deforestation?

5. What energy source has been linked to deaths of migratory birds?

6. What energy source can only be used in a few places in the world?

7. What energy source can be utilized in areas with volcanic activity?

8. What energy source destroys landscapes by flooding them?

Free Response Questions

A family is considering using solar energy to heat and produce electricity for their house and to help lower their electricity bills and lower their impact on global warming.

1. Describe both active and passive methods the family could use to reach their goal.

2. What are some of the challenges the family will run into by using solar energy?

3. What are some of the benefits the family will enjoy by using solar energy?

4. What benefits has the family done for the environment by making the move to solar energy?

Multiple Choice Answers and Explanations

1. B. Geothermal energy is generated from heat deep within the earth. It is not affected by the Sun.

2. D. When a river is dammed, the river ecosystem is destroyed and fish are no longer able to migrate upstream.

3. D. Oil supplies most of the world's commercial energy.

4. C. Biomass leads to deforestation. People cut down the forest to plant crops that are then burned as fuel, or they use the trees as fuel.

5. B. A concern about windmills is that migratory birds are attracted to them and get killed in the blades of the windmill.

6. E. Only 30 places in the world have tides large enough to utilize tidal power effectively.

7. D. Geothermal energy is generated by capturing heat from deep within the earth. This heat can be captured in areas where volcanic magma is close to the surface.

8. A. Dams flood areas above the dam, which destroys the natural ecosystem and can reduce the area's biodiversity.

Free Response Answers

1. Photovoltaic cells, pumps, or other technologies that generate heat and electricity from the Sun are all examples of active solar systems. Adding shades to windows, installing a south-facing window, or adding solar panels to a house are all examples of passive solar systems.

2. The sun is only out during the day, so the family would have to consider how to generate heat and electricity at night, during long winter months, and when the sky is overcast. These challenges can be solved by using batteries to store the energy or by being connected to the city's electrical grid and using it when necessary.

3. The family will enjoy a lower electricity bill. They also might generate enough power by using solar panels to sell their electricity back to the grid.

4. The family has helped the environment by reducing our nation's contribution to greenhouse gas emissions and dependence on foreign oil. The family also has lessened the damage that coal mining does to the environment and the acid rain that mining generates.

Environmental Hazards and Human Health

15.1 Links between Human Health and the Environment

Hazards are anything that can cause

- injury, disease, or death to humans.
- damage to personal or public property.
- deterioration or destruction of environmental components.

Risk is the probability of suffering injury, disease, death, or loss as a result of a hazard.

Vulnerability is how susceptible a person is to a certain risk. The poor are more susceptible to certain risks than are others.

Risk = Hazard x Vulnerability

Morbidity	The incidence of *disease* in a population
Mortality	The incidence of *death* in a population
Epidemiology	The study of disease in populations

Examples of Cultural Hazards	**Examples of Biological Hazards**
Eating too much Driving too fast Smoking/drugs/alcohol Risky sexual practices Hazardous occupations Living in inner cities	Black plague Typhus Small pox Diarrhea Tuberculosis Malaria
Examples of Physical Hazards	**Examples of Chemical Hazards**
Natural disasters Hurricanes Tornadoes Floods Forest fires Landslides Volcanic eruptions	Cleaning agents Pesticides Fuels Paints Medicines Carcinogens

15.2 Pathways of Risk

The World Health Organization states that poverty is the world's largest killer. The poor

- have less than adequate health care.
- lack clean drinking water.
- lack nutritious food.
- lack clean air, sanitation, and shelter.

The world's number one risk factor is being underweight, particularly in children.

Education and advances in socioeconomic development are ways to change this factor.

<u>Facts about smoking</u>

- Tobacco use is the leading cause of preventable deaths in the United States and ranks fourth globally.
- About 21% of Americans smoke, and half of these people will die or become disabled because of smoking.
- Tobacco is the only known legal product that kills half of its users.
- The World Health Organization states that 5 million smokers die each year from smoking-related causes.
- The EPA classifies secondhand smoke as a carcinogen.

<u>Terms related to risk</u>

- **Chronic** refers to symptoms that show up over a period of years.
- **Acute** refers to life-threatening reactions that occur within hours or days.
- **Carcinogenic** is a pollutant that leads to cancer.

Indoor air pollution

- Secondhand smoke is a form of indoor pollution.

- Biofuels like wood and animal dung for cooking and heat are also forms of indoor pollution

- Indoor pollutants are dangerous because they can expose people to high levels of toxic substances, like carbon monoxide and soot.

- Asthma is a common consequence of indoor pollution and is at epidemic proportions in the U.S.

Toxicology is the study of toxic substances on human health.

15.3 Risk Assessment

A **risk assessment** is a process of evaluating the risks of a hazard.

Risk assessment has four steps:

1. Hazard assessment
2. Dose-response assessment
3. Exposure assessment
4. Risk characterization

One **DALY**, or disability-adjusted life year, represents the loss of one healthy year of a person's life.

Sample Multiple Choice Questions:

1. Anything that causes injury, disease or death to humans, damage to personal or public property, or deterioration or destruction of environmental components is called

A. risk.
B. morbidity.
C. hazard.
D. vulnerability.
E. epidemiology.

2. The incidence of disease in temperate climates has decreased. This decline is best described as a decrease in

A. mortality.
B. morbidity.
C. carcinogens.
D. climate.
E. nitrification.

3. Earthquakes are best described as which of the following types of hazard?

A. physical
B. cultural
C. chemical
D. biological
E. aesthetic

4. What is the number one risk factor for determining mortality?

A. educational attainment
B. malnourishment
C. climate change
D. deforestation
E. eutrophication

5. What is the leading cause of preventable death in the United States?

A. alcohol consumption
B. tuberculosis
C. breast cancer
D. tobacco use
E. pesticide exposure

6. Factors that contribute to the increase in world hunger include all of the following *except*

A. poverty.
B. land use.
C. climate change.
D. population growth.
E. fertilizer use.

7. What is radon gas accumulation in basements most likely to cause?

A. acute illness
B. chronic illness
C. cataracts
D. diverticulitis
E. hair loss

Free Response Questions

Matamoras, Mexico, has an abnormally high incidence of children born with part of their brain missing. Epidemiologists studying the area would want to consider many risk factors and hazards to which the children or their mothers have been exposed.

1. What risk factors would the epidemiologists need to consider when determining the cause of these illnesses?

2. Why are the environmental issues causing problems in children rather than adults?

Multiple Choice Answers and Explanations

1. C. A hazard is anything that can cause, injury, disease or death to humans, damage to personal or public property, or deterioration or destruction of environmental components.

2. B. Morbidity is the incidence of disease in a population.

3. A. Earthquakes are an example of a physical hazard. Other examples include natural disasters, hurricanes, tornadoes, floods, forest fires, landslides, and volcanic eruptions.

4. B. Malnourishment (hunger) is the number one risk factor, especially in children.

5. D. Tobacco use is the leading cause of preventable death in the United States and ranks fourth globally.

6. E. Fertilizer use is not a factor that contributes to the increase in world hunger.

7. B. Long-term radon exposure causes chronic illness such as cancer.

Free Response Answers

1. The epidemiologists would need to consider chemical, physical, cultural, and/or biological factors when studying these children. They would want to find the cause of the birth defect and prevent the hazard from getting into the drinking water and food supply.

2. Children are more susceptibility to diseases and environmental hazards than adult because of children's smaller size, weaker immune system, and immature development.

Pests and Pest Control

16.1 The Need for Pest Control

A **pest** is any organism that is harmful or destructive or interferes with humans or our social or economic endeavors.

An **herbicide** is a chemical used to kill plants.

A **pesticide** is a chemical used to kill animals and insects.

Pest control is important for two reasons:

- It prevents diseases such as malaria.
- It destroys pests that harm crops.

Integrated pest management (IPM) is an approach that uses both chemical (pesticides) and ecological (protecting people or animals rather than eradicating the pest) methods to control pests.

LD50 is the amount of a material, given all at once, which causes the death of 50% (one half) of a group of test animals. LD50 is one way to measure the short-term poisoning potential (acute toxicity) of a material. LD stands for *lethal dose*.

16.2 Promises and Problems of the Chemical Approach

DDT, which was discovered in the 1930s, initially seemed like the perfect pesticide. It killed many different kinds of insects and, because it stayed in the environment for a long time, worked long after you sprayed it.

Pros of DDT

- DDT saved many peoples lives. During World War II, it was used to control the spread of body lice and, consequently, deaths from typhus fever. It also has been used successfully in the tropics to stop the spread of malaria.
- By controlling pests, DDT also increased crop yields.

Cons of DDT

- Overtime, pests became resistant to DDT, and larger amounts of DDT had to be used to get the same effect.
- Insects that were not a problem before became a problem when DDT killed off their natural predators.
- Once pests became resistant to DDT, their populations exploded.
- The environment was harmed because of DDT, especially bird populations.

There are human health concerns related to pesticide use.:

- Pesticides have been linked to cancer.
- Birth defects and infertility have been linked to people exposed to pesticides.

There are three main environmental concerns related to pesticide use.

- **Bioaccumulation** is the process of accumulating higher and higher doses of a pesticide up the food chain. This process happened with DDT, and it was especially harmful to predatory birds. Figure 16–9 shows that phytoplankton, which is at the bottom of the food chain, get a very small dose of DDT (0.04 ppm). As the zooplankton eat the phytoplankton and as the fish eat the zooplankton and as the osprey eat the fish, the concentration of DDT gets larger (2.0 ppm). By the time the bald eagle eats the osprey, the DDT concentration has reached toxic levels (20 ppm). In the case of DDT, this high concentration of DDT made the eggs of the birds weak, which lowered their reproduction rate to near extinction.

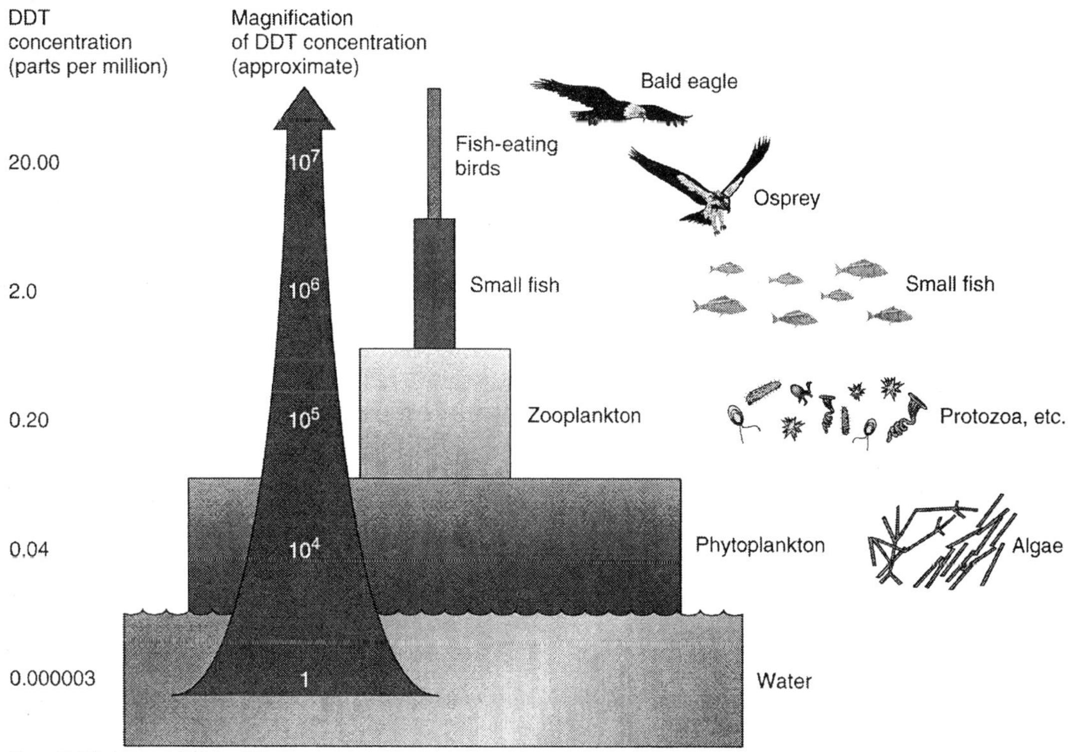

- Figure 16-9 Environmental Science, 10/e
© 2008 Pearson Prentice Hall, Inc.

- **Biomagnification** is the multiplying effect of bioaccumulation in a food chain.

- **Nonpersistant pesticides** are pesticides that break down into simple nontoxic substance within a few weeks of being used. However, nonpersistent pesticides are very toxic and pose a serious danger to humans exposed to them.

16.3 Alternative Pest Control Methods

One alternative pest control method is **cultural control.** With this method, the environment is altered in such a manner that pests cannot thrive. Changing bed linens to prevent bedbugs or cleaning the house to prevent roaches are both examples of cultural controls.

Another method is to use **natural enemies,** or predator bugs, to control pests. For example, ladybugs can be purchased at most garden stores to predate on aphids, a common garden pest.

Another method is **genetic control.** This means breeding plant varieties that are resistant to pests. For example, plants can be breed to produce **chemical barriers** that kill or repulse pests. Another method of genetic control is to release a group of **sterilized males** into the environment. These sterilized males mate with females, which prevents future attempts for the female to mate.

Pheromones are chemicals that a species emits to attract members of the same species. Scientists use these pheromones to trap pests in devices baited with poison.

16.4 Socioeconomic Issues in Pest Management

The **economic threshold** is a measure used to determine when the damage caused by pests is considerable and outweighs the cost of applying a pesticide.

Insurance spraying refers to the use of pesticides to prevent a pest before it becomes a problem.

Cosmetic spraying refers to the use of pesticides to control pests that make fruits or vegetables look bad and less desirable to consumers.

Organic foods are grown with little or no pesticides, chemical fertilizers, antibiotics, or hormones. Organic farming produces smaller yields than nonorganic farming methods. However, sales of organic foods have increased by 20% each year for the past 10 years.

16.5 Pesticides and Policy

FIFRA (Federal Insecticide, Fungicide and Rodenticide Act of 1947 and amended in 1972) requires manufacturers to register pesticides with the EPA before marketing them and to test their toxicity levels.

The **FQPA** (Food Quality Protection Act of 1996) set standards for pesticides applied to foods (and special concern for children). Among other things, the act requires that

- a pesticide be prohibited from being used if it is linked to cancer.
- all possible exposures (not just food) to pesticides be evaluated.
- older pesticides be reevaluated to see if they meet the new standards.
- the standards apply to both processed and raw foods.

Tolerance is the limit to the amount of pesticide that can remain on food.

Multiple Choice Questions

1. Bald eagles were almost brought to extinction because levels of DDT were found to cause their eggs to be weak and easily broken. How did DDT reach these levels?

A. Chemicals used in one location permeated throughout the biosphere, causing damage far away from the initial application.
B. DDT was applied extensively to help control the bald eagle population.
C. DDT evaporates and later condenses near bald eagle habitats.
D. DDT used to control insect populations spread up the food chain through the process of bioaccumulation.
E. Bald eagles outcompeted other predators and depleted their natural source of food.

2. LD50 is best defined as

A. the amount of pesticide necessary to displace 50 liters of denatured alcohol
B. the amount of pesticide in a 50 molar solution of sulfuric acid
C. the amount of pesticide necessary to kill 50% of an animal test group
D. the number of years necessary for a pesticide to break down into nonlethal components
E. the London-Dubai agreement to ban pesticide use by 2050

3. If the LD50 level of a pesticide for rats is 200 milligrams per kilogram of body mass, then this means that

A. 50 rats would die if exposed to this pesticide
B. 50 rats would get sick if exposed to this pesticide
C. 50 percent of all people exposed to this pesticide would die
D. 50 percent of all people exposed to this pesticide would get sick
E. 100 out of every 200 rats would die if exposed to this pesticide

4. Breeding plants to be pest resistant is an example of

A. chemical barriers
B. genetic controls
C. pheromones
D. cultural controls
E. a form of pest control that is still a long way from being used

5. Which law requires that every bottle of pesticide list its ingredients, environmental hazards, and health risks?

A. FQPA
B. FIFRA
C. Clean Water Act
D. Clean Air Act
E. RCRA

6. Some pesticides will magnify their way up a food chain. If a certain pesticide were found to be 2 ppm in small fish, what would the amount be in larger fish that eat the small fish?

A. 2 ppm
B. .2 ppm
C. .02 ppm
D. 20 ppm
E. 200 ppm

7. Ladybugs are sold at garden supply stores

A. as pollinators for plants
B. to breed, repopulate, and control weeds
C. as an organic method of controlling soil erosion
D. as a natural control for certain insects
E. to increase fertility

8. Rachel Carson wrote *Silent Spring*, a book about the effects of DDT and other pesticides on the environment. Rachel Carson was most concerned about

A. biomagnification
B. global climate change
C. nonpersistent pesticides
D. soil denitrification
E. the effects of pesticides on organic farming

9. Pesticides have been linked to all of the following human health concerns *except*

A. breast cancer
B. birth defects
C. infertility
D. defective sperm counts
E. tuberculosis

10. Pesticides are beneficial for all of the following reasons *except*

A. pesticides eliminate pests that spread disease
B. pesticides add important nutrients to the soil
C. pesticides keep fruit and vegetables aesthetically appealing to consumers
D. pesticides kill pests that harm crops
E. pesticides allow people to live more comfortably

Free Response Questions

1. Many farmers have switched from using only pesticides to integrated pest management (IPM) techniques. List the characteristics of IPM, and describe its positive environmental effects.

Multiple Choice Answers and Explanations

1. D. DDT reached the level of bald eagles through bioaccumulation, the process by which a chemical accumulates higher and higher doses as it moves up the food chain.

2. C. LD50 is the amount of a material, given all at once, which causes the death of 50% (one half) of a group of test animals.

3. E. LD50 is the amount of a material, given all at once, which causes the death of 50% (one half) of a group of test animals. So if the LD50 level for rats is 200 milligrams per kilogram of body mass, then 100 out of every 200 rats would die if exposed to the pesticide.

4. B. Genetic controls are used to breed pest-resistant plant varieties.

5. B. FIFRA, or the Federal Insecticide, Fungicide and Rodenticide Act, requires that manufacturers register pesticides with the EPA before marketing them and list their ingredients and the environmental hazards and health risks associated with them.

6. D. If a certain pesticide were found to be 2 ppm in small fish, then the amount of pesticide found in the larger fish would be 20 ppm. It is assumed that toxins magnify by a factor of 10 as you move up each level in a food chain.

7. D. Ladybugs are predator bugs that, when released, help control aphid populations.

8. A. Rachel Carson's book discussed how DDT was moving up the food chain and harming birds. Thus, she was concerned with biomagnification.

9. E. Pesticides have been linked to cancer, low sperm count, birth defects and infertility to name a few. They have not been linked to tuberculosis.

10. B. Pesticides are beneficial for all the reasons listed except adding important nutrients to the soil.

Free Response Answers

1. Integrated pest management (IPM) is an approach to controlling pests that uses both chemical and ecological methods to bring about long-term management of pest populations with a minimal environmental impact. IPM is used in places where health risks or the cost of pesticides prevent large usage. IPM takes into account when the pest is active, why the pest is there, how the pest arrived, and what parasite or predator could be used to control the pest. By switching from total pesticide use to IPM, the amount of residual pesticide in the environment would decrease. It would reduce the human health effects related to pesticide use and the amount of fuel needed to apply pesticides to crops. Soil and water contamination would decrease, and pests would be less likely to build a resistance to the pesticides.

Water Pollution and Its Prevention

17.1 Water Pollution

Point source pollution occurs when factories, power plants, mining operations, and others discharge pollutants directly into a body of water. With this type of pollution, it is easy to identify the source and to fix the problem. (You can "point" to whom is polluting.)

Nonpoint source pollution is runoff from various sources such as city streets, rural homes, cropland, and animal feedlots. A little bit of pollution from everyone adds up to a lot of pollution. This type of pollution is harder to fix and to control. (Since no single source is polluting, it is difficult to "point" to the problem.)

There are five different types of water pollutants:

- **Pathogens** are infectious agents that cause sickness and death. Pathogens originate from humans or other animals carrying the disease.

- **Organic wastes** are organic matter (leaves, grass clippings, trash) that can stimulate plant growth. Organic wastes can lower the amount of dissolved oxygen in a body of water.

- **Chemical pollutants** are water-soluble inorganic chemicals like heavy metals (lead, mercury, arsenic, salt, oil, pesticides, detergents) that are toxic to water.

- **Sediment** from erosion and weathering can end up in the water. Sediments can block light from entering a body of water, kill animals by clogging their gills, and bury organisms that live at the bottom of the water.

- **Nutrients** such as nitrogen and phosphorous get into the water and encourage the growth of plants. These nutrients can come from fertilizers, sewage, or other sources that runoff into the water and lead to eutrophication.

17.2 Wastewater Management and Treatment

Storm drains collect water from runoff rain.

Sanitary sewers collect water from sinks, tubs, and toilets in people's homes.

There are four methods for removing pollutants from wastewater. Each method is summarized in the chart below and illustrated in Figure 17–9.

Preliminary Treatment (Bar screens)	This method removes debris and grit by using screens and settling of the grit.
Primary Treatment (Grit-settling tank)	This method removes particulate organic material by allowing it to settle to the bottom and pouring off the water.
Secondary Treatment (Activated-sludge system)	This method removes organic material by using organisms (natural decomposers and detritus feeders) to "eat" the organic material. Oxygen is pumped in to encourage the organic material to grow.
Final Cleansing and Disinfecting (Chlorination tank)	This method uses an agent such as chlorine gas to disinfect the water and then another agent to dechlorinate the water so it is safe to put into a waterway.

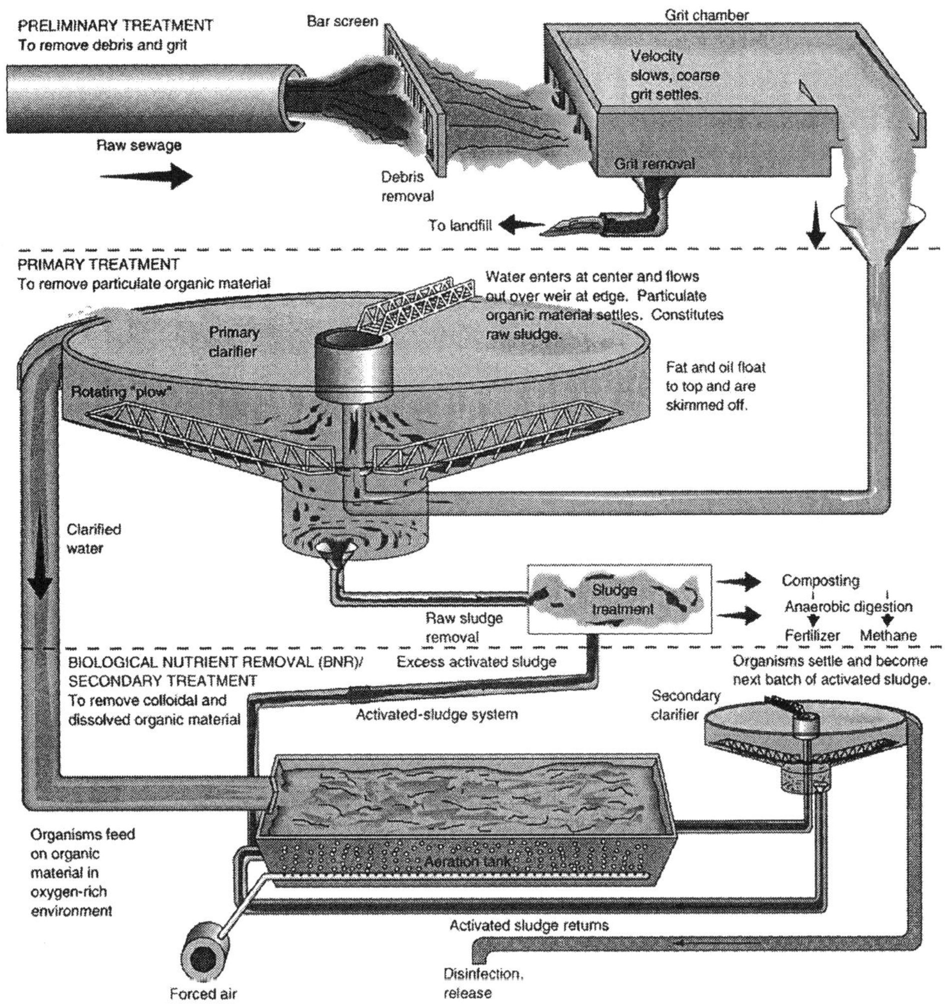

Figure 17-9 Environmental Science, 10/e
© 2008 Pearson Prentice Hall, Inc.

There are three methods for treating **sludge**, the solid waste that settles out of the sewage water after primary treatment.

- The **anaerobic digestion** method allows bacteria to feed on the sludge. The decomposition produces a biogas (which is mostly methane) that can be burned as fuel.

- **Composting** is when sludge is mixed with wood chips and then left to be broken down by bacteria and other decomposers. The resulting organic material is a rich humus-like material that can be used to supplement poor soils.

- **Pasteurization** is when heat is used to kill pathogens in sludge. Heating causes the sludge to pellet. These pellets are used as fertilizers.

There are three alternative treatment systems:

- With **on-site wastewater treatment systems** (e.g., septic tanks and leaching fields), wastewater flows into a tank, organic material settles to the bottom, and the water flows into the soil. This water is clean enough to be used on vegetable gardens.

- Wastewater that has gone through the secondary treatment process can be used to **irrigate** areas of land such as golf courses, croplands, and parks.

- Another alternative is to **reconstruct wetlands,** which are naturally capable of cleaning runoff and other wastewaters before they enter larger bodies of water.

17.3 Eutrophication

There are two different kinds of aquatic plants. **Benthic plants** are plants attached to the bottom of a body of water. **Phytoplankton** includes algae, protists, and photosynthesizing bacteria.

Aquatic plants can be overnourished (i.e., have too high a level of nutrients). Too many nutrients can result in a series of changes in the water called **eutrophication**. Here are the changes (be sure you understand this process):

- As nutrients enter a body of water, the aquatic plants living in the water become fertilized.

- This fertilization causes phytoplankton to multiply.

- As phytoplankton multiply, they reach a maximum population and then begin to die off.

- The dead phytoplankton settle to the bottom of the water and become food for decomposers (mainly bacteria).

- The explosive growth of bacteria that follow consume a lot of oxygen, causing depletion of oxygen in the water.

- Hypoxic conditions (no dissolved oxygen) result with fish and other aquatic animals dying.

If this process is natural, it is called *natural eutrophication*. If it is the result of human activity, it is called *cultural eutrophication*.

There are two ways to combat eutrophication:

- You can **attack the growth of vegetation** by putting herbicides in the water, using aeration, harvesting aquatic weeds, or drawing water down so rooted plants die.

- Or you can **stop nutrients from being added to the water** by identifying point or nonpoint source pollutants and then banning them or educating people about better practices for fertilizers and other chemicals.

17.4 Public Policy

Four legislative acts have been enacted to protect the nation's waters. They are summarized in the chart below.

The Clean Water Act of 1972	This act and its amendments require permits for any discharge of pollution, strengthen water quality standards, and encourage the use of pollution-control technology.
Marine Protective Research and Sanctuaries Act of 1972	This act prevents unacceptable dumping in oceans.
Safe Drinking Water Act of 1996	This act authorizes the EPA to regulate the quality and safety of public drinking water supplies, maintain drinking water standards for contaminants, and set requirements for chemical and physical treatment of drinking water.
Water Quality Act of 1987	This act amended the Clean Water Act and created a loan to support construction of treatment plants, addressed regional pollution, and addressed nonpoint source pollution (the nation's leading water pollution problem).

Multiple Choice Questions

1. All of the following can be used to treat wastewater *except*

A. chlorine gas.
B. ozone.
C. ultraviolet light.
D. Clorox®.
E. alcohol.

2. Animals can add fecal coliform bacteria to a body of water. This has all of the following effects *except*

A. causes eutrophication.
B. fouls the water and spreads disease.
C. causes toxic metals to precipitate out of a solution.
D. causes the level of dissolved oxygen in the water to drop.
E. causes a bloom of phytoplankton.

3. If wastewater is not treated properly, then

A. nitrates and phosphates get into the water.
B. excessive methane builds up.
C. fires can result .
D. it leads to ozone depletion.
E. it increases the incidence of cystic fibrosis.

4. Which step in the treatment of wastewater uses bacteria or other microorganisms to consume the organic material?

A. bar screens
B. primary treatment
C. activated-sludge system
D. final cleansing and disinfecting
E. chlorination tank

5. Cultural eutrophication is caused by

A. natural organic materials getting into a body of water.
B. cyanobacteria living in the water and reproducing.
C. gamma radiation .
D. volcanic eruptions.
E. fertilizers from farms running off into a body of water.

6. Which step in the treatment of wastewater uses chemicals like sulfur dioxide to neutralize the effect of chlorine gas?

A. bar screens
B. primary treatment
C. activated-sludge system
D. final cleansing and disinfection
E. chlorination tank

7. Of the following, which is the best example of a nonpoint source of water pollution?

A. factory effluents
B. raw sewage from a storm drain
C. oil spill
D. power plant smokestack
E. agricultural runoff

8. Many things can be done safely with wastewater from homes, including

A. washing clothes.
B. watering livestock.
C. watering city parks and golf courses.
D. filling swimming pools.
E. bathing.

Sample Free Response Questions:

1. Look at the following diagram of a sewage treatment plant. Identify and describe the function of any three parts labeled on the diagram.

Multiple Choice Answers and Explanations

1. E. All of the options listed except alcohol can be used to treat wastewater.

2. C. When fecal coliform is added to a body of water, it can result in all of the options listed except causing toxic metals to precipitate of a solution.

3. A. Nitrates and phosphates are wastes that come from untreated water.

4. C. During secondary treatment, organisms (natural decomposers and detritus feeders) "eat" the organic material, including pathogens. The activated-sludge system is the most common secondary treatment system.

5. E. Cultural eutrophication results from human activities like fertilizers from agricultural lands running off into a body of water

6. D. In the final cleansing and disinfection step, chlorine is added to disinfect the water and then another agent like sulfur dioxide is used to neutralize the chlorine.

7. E. Nonpoint sources are pollutants that cannot be attributed to any one source. Agriculture runoff comes from many different farms.

8. C. The only safe thing that is being done with wastewater is using it to water public parks and golf courses.

Free Response Answers

1. See diagram below.

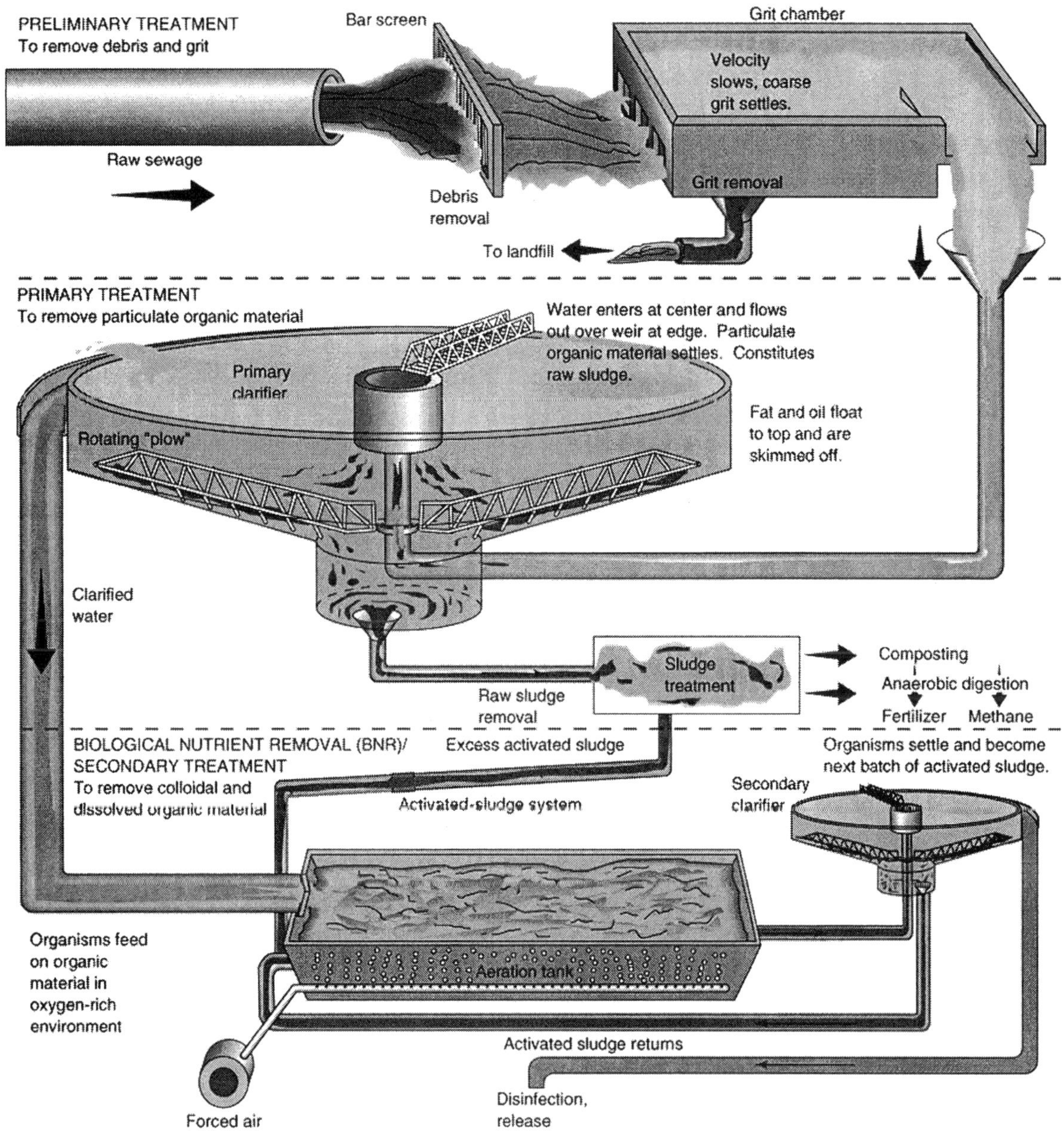

Figure 17-9 Environmental Science, 10/e
© 2008 Pearson Prentice Hall, Inc.

Municipal Solid Waste: Disposal and Recovery

18.1 The Solid-Waste Problem

Municipal solid waste refers to trash, refuse, or garbage from homes and small businesses.

In the past, most municipal solid waste was burned in open dumps. This practice led to air pollution and rat and fly infestations, among other problems. Later, open dumps were turned into landfills.

With a **landfill,** waste is put into the ground and covered with at least 6 inches of earth. This practice does not pollute the air or attract rats and flies. However, there are several problems associated with landfills.

- The most serious problem is **leachate generation and groundwater contamination**. When it rains, water seeps down through the trash and picks up pollutants. If the leachate (water with pollutants) is not dealt with properly, it will seep through the ground and into the groundwater.

- Landfills can also produce large amounts of **methane gas**. As the trash decomposes, biogas is produced, which is about two-thirds methane. This methane can seep through the soil and rock and end up in basements, where it can cause fires or explosions. Many landfills, however, are now using this methane as a source of fuel.

- Another problem associated with landfills is **incomplete decomposition**. Many things put into a landfill either do not decompose or decompose slowly. Even paper and wood will not decompose in a landfill when there is not enough water to help with decomposition.

- **Settling** is another problem. As waste settles, it compacts and decomposes, which can cause the land to sink and collect water.

There are several ways to improve landfills:

- You can put landfills on high ground above a water table.

- You can build leachate collection systems to pump water out and treat it before it enters the groundwater.

- You can stack the trash in a pyramid shape so water will run off it rather than run into the landfill.

- You can put groundwater monitoring wells in place to make sure the landfill is not leaking into the groundwater.

Some states burn waste and use it to create energy. There are three main advantages of combustion:

- Combustion reduces the amount of waste.
- The ash left over after combustion can be captured and put back into a landfill or used in construction.
- Using waste as a fuel for electricity production reduces our nation's need for coal.

The disadvantages of combustion are as follows:

- Combustion produces air pollution and odors.
- Combustion facilities are expensive to build.
- The ash left over after combustion may contain toxic materials.

18.2 Solutions to the Solid-Waste Problem

The following are examples of things individuals can do to reduce the amount or toxicity of trash created (**source reduction**):

- You can reduce the weight of items.
- You can use computers to lessen the amount of paper used.
- You can make reusable items.
- You can make items last longer so they can be reused.
- You can get off bulk mailing lists.
- You can compost yard waste.

Recycling is another way to address the solid-waste problem. Recycling is a great option for the following reasons:

- It uses material from one object and makes it into something new.
- It saves energy and resources.
- It helps to decreases pollution.

There are many items that can be recycled. They include

- paper (the most important is newspaper because so much is generated),
- glass,
- plastic,
- metals,
- yard wastes,
- textiles,
- old tires.

18.3 Public Policy and Waste Management

Four federal legislative acts address waste management. Each is summarized in the following chart.

Act	Summary
Solid Waste Disposal Act of 1965	This act gave jurisdiction of solid waste to the EPA.
Resource Recovery Act of 1970	This act directed attention to recycling programs.
Resource Conservation and Recovery Act (RCRA) of 1976	This act closed local dumps and set regulations for landfills.
Superfund Act of 1980	This act addressed abandoned hazardous-waste sites.

Integrated waste management is a system that uses several approaches to deal with our nation's solid waste problems. It focuses on both waste reduction (lowering the amount of waste we produce and dispose) and recycling and reuse. Some even argue for banning the disposal of recyclable items in landfills.

Multiple Choice Answers and Explanations

1. All of the following are problems associated with landfills *except*

A. leachate seeping into groundwater.
B. incomplete decomposition.
C. rats and flies.
D. methane production.
E. settling.

2. The largest component of waste in the United States is

A. glass.
B. plastic.
C. paper.
D. organic material.
E. aluminum.

3. In the United States, most of our trash is

A. incinerated.
B. put into landfills.
C. recycled.
D. put into drums and stored.
E. used to generate electricity.

4. Which of the following is not a problem associated with incineration?

A. It causes air pollution.
B. It reduces the amount of waste that ends up in landfills.
C. It uses a lot of energy.
D. Toxic fumes can be released into the environment.
E. The ash may contain hazardous material.

5. Recycling our trash rather than putting it into landfills has many benefits *except*

A. it uses less raw materials and generates less waste.
B. it decreases pollution.
C. it produces toxic leachate.
D. it saves energy and resources.
E. it makes something new out of old material.

6. Integrated waste management is a system that

A. uses the processes of reduce, reuse, and recycle.
B. encourages producers to overpackage materials.
C. increases the amount of air pollutants.
D. uses more energy than traditional methods.
E. increases landfill volume.

7. Many changes have been made to make disposing of municipal waste safer. One of these changes is

A. digging down below the water table to put our solid waste.
B. using a liner system to catch and hold water that gets into landfills.
C. placing landfills near rivers to reduce transportation costs.
D. stacking landfills in a pyramid shape to help with water runoff.
E. capping landfills so methane gas cannot escape.

8. Waste-to-energy generators are

A. used by individual households to smash and compact garbage.
B. a system of curbside recycling for homeowners.
C. a system where methane generated at a landfill is used for fuel.
D. not used because they increase fuel costs.
E. growing to become the largest energy producer around the world.

Free Response Questions

1. What are the similarities and differences between a sanitary landfill and an unregulated garbage dump?

2. List several obstacles to recycling in the United States, and suggest ways to overcome these obstacles.

3. What environmental and human health risks are associated with landfills?

Multiple Choice Answers and Explanations

1. C. Rat and fly infestations are diminished when sanitary landfills are built.

2. C. Paper is the largest component of waste in the United States, which explains why it is so important to recycle it.

3. B. Most trash ends up in a landfill.

4. B. Incinerators do reduce the amount of waste that ends up in landfills. However, this is a benefit, not a problem.

5. C. Leachate is not a benefit of recycling trash. Leachate is water that becomes polluted as it passes through a landfill.

6. A. Integrated waste management is a system that uses several approaches to address waste management, such as reduce, reuse, and recycle.

7. D. Pyramid-shaped landfills help keep water from seeping into a landfill by directing its runoff.

8. C. Waste-to-energy generators are used to make electricity out of the methane produced by the decomposition in landfills.

Free Response Answers

1. Unregulated garbage dumps produce clouds of smoke, bad smells, and air pollution and also are breeding grounds for flies and rats. Sanitary landfills control each of these. Both accept trash from homes and businesses and both can produce leachate and water problems, produce methane gas, and have problems with settling and improper decomposition.

2. There are several obstacles to recycling. Many times, it is more expensive to recycle than to buy new material. Also, some people are unwilling to separate out their trash from their recyclables. There are ways to overcome these obstacles. By increasing the demand for recycled products, the cost of making these products can be lowered. Cities can make recycling mandatory and fine people for noncompliance.

3. The environmental risks associated with landfills include toxic material getting into the groundwater, pollutants being released into the atmosphere, and soil being contaminated. The human health risks include pollutants entering the drinking water and methane gas causing explosions.

Hazardous Chemicals: Pollution and Prevention

19.1 Toxicology and Chemical Hazards

Toxicology is the study of harmful effects of chemicals on human and environmental health.

Hazardous Materials (HAZMATs) are chemicals that present a hazard or a risk. A chemical is categorized as a HAZMAT if it has one of the following properties:

- It is ignitable. Will it catch fire?

- It is corrosive. Will it corrode storage tanks and equipment?

- It is reactive. Does it explode or create toxic fumes when mixed with water?

- It is toxic. Does it make people ill if they ingest or inhale it?

HAZMATs may enter the environment in several ways, including through production processes, pesticides, fertilizers, and leaks from storage tanks.

Two types of toxic chemicals do not degrade easily in the environment:

- **heavy metals** such as lead, mercury, arsenic, cadmium, tin, chromium, zinc and copper that are used in industry, particularly in making batteries and electronics. (neurotoxins)

- **synthetic organics** such as plastics, synthetic fibers, synthetic rubber, paint coatings, solvents, pesticides, wood preservatives, and hundreds of others. (Chlorinated Hydrocarbons ex. DDT)

19.2 A History of Mismanagement

In the past, chemical wastes were burned. The resulting fumes went into the air, were flushed away into waterways, or were vented directly into the air. Human health problems resulted as a consequence.

Beginning in the 1970s, several land methods were developed for disposing of these chemicals.

- With **deep-well injection,** waste is injected into a hole that is drilled to a depth thousands of feet below the water table and then sealed off to prevent back up.

- With **surface impoundments,** a "pond" is excavated and filled with liquid wastes. For example, leachate is drained out of a landfill and then pumped into a surface impoundment (called a leachate collection tank). After the chemical settles to the bottom and the water evaporates, the waste is put back into the landfill.

- When placed in **landfills,** hazardous wastes are put into drums and then treated according to rigid EPA standards. Only 23 landfills are licensed to receive hazardous wastes in North America.

- **Midnight dumping and orphan sites** refer to illegally dumping hazardous materials under the cover of darkness (midnight dumping) or on abandoned property (orphan sites).

Love Canal, near Niagara Falls in New York, is an example of what can happen when disposal of hazardous materials is unregulated. Hooker Chemical and Plastics Company bought an abandoned canal (Love Canal), filled it with hazardous wastes, capped it with a clay cover, and then sold the area to the local school board. After several years, the surface of the dump began to collapse, exposing barrels of chemical wastes and releasing fumes and chemicals into the surrounding environment. People in the area of Love Canal began reporting serious health problems, including birth defects and miscarriages. Millions of dollars were spent cleaning up the site and on lawsuits.

CERCLA (Comprehensive Environmental Response, Compensation, and Liability Act) of 1980, also known as **Superfund**, is legislation that provides a trust fund for identifying and cleaning up chemical waste sites.

19.3 Cleaning Up the Mess

Groundwater remediation is a method of cleaning groundwater that has been contaminated by hazardous materials. Wells are drilled, and the contaminated groundwater is pumped out, purified, and then put back into the ground or discharged into surface waters.

Brownfields are areas that are not as bad as Superfund sites but that still need to be cleaned and reclaimed. These are from abandoned industrial/commercial facilities. The Brownfield Act of 2002 provides grants for assessment and remediation of sites.

Gasoline stations store gasoline in **underground storage tanks (USTs)** to help prevent fire and explosions. These tanks, however, last only 20 years before they begin leaking. Federal legislation now requires that these tanks be monitored regularly and remediated within 72 hours if a leak is detected.

19.4 Managing Current Hazardous Wastes

Any firm discharging more than a certain volume into natural waterways must have a **discharge permit.** The CWA (Clean Water Act) controls this.

The **Resource Conservation and Recovery Act (RCRA)** was passed in 1976 to help control the land disposal of hazardous wastes. The RCRA established the following three requirements:

- All land disposal facilities must have a permit;

- Toxic wastes must be pretreated to prevent leaching;

- Toxic substances being transported must be well documented and ready for inspection at any time.

The transportation of hazardous materials also must meet certain **Department of Transportation Regulations (DOT Regs)**. These regulations are designed to reduce the risk of spills, fires, and poisonous fumes. Any vehicle carrying hazardous waste must

- identify on large placards the hazardous material they are carrying (e.g., flammable, corrosive, toxic, etc.);

- take designated routes on certain highways or during certain times.

The **Occupational Safety and Health Act (OSHA)** protects workers who are exposed to hazardous materials. According to OSHA, workers must have information about the hazardous materials they are working with and about the protective equipment they should be wearing. Some of this information is stored on **material safety data sheets (MSDS)**, which accompany more than 600 chemicals at all times.

The **Emergency Planning and Community Right-to-know Act (EPCRA)** addresses accidents of hazardous wastes. According EPCRA, all companies handling 5 tons of hazardous material must give information about the material to a local emergency planning committee. The committee consists of fire and police departments, hospitals, and other emergency responders that may be involved in case of an emergency. The committee uses the information to draw up accident scenarios and plans for responding to accidents.

19.5 Broader Issues

Environmental justice refers to the fair and equal treatment of all people regardless of race, color, national origin, or income. No group should bear a disproportionate share of the negative environmental consequences resulting from the use, disposal, and regulation of hazardous materials. However, in reality, waste sites are often located in non-Caucasian and low-income neighborhoods. This practice is referred to as **environmental racism**.

Pollution control refers to practices designed to prevent pollution from entering an environment. A catalytic converter, a device that filters a car's exhaust, is an example of a pollution control.

Pollution prevention refers to practices designed to eliminate or reduce the amount of pollution being produced. Hybrid cars are an example of pollution prevention.

Multiple Choice Questions

1. What law states that all hazardous materials must be pretreated and disposed of properly with the correct permits?

A. OSHA
B. RCRA
C. DOT Regs
D. Clean Water Act
E. Clean Air Act

2. What law states that all workers must be provided with hazardous material sheets that inform them of the chemicals with which they are working?

A. OSHA
B. RCRA
C. DOT Regs
D. Clean Water Act
E. Clean Air Act

3. Citizens led by Lois Gibbs demanded that something be done about the contamination at Love Canal near Niagara Falls. The problems that developed were a result of

A. a nuclear power plant leaking radiation.
B. point-source pollution from a sewage treatment plant contaminating the local river and water supply.
C. contaminated soil from local farmers using pesticides.
D. contaminated land from a company burying toxic wastes.
E. acid rain from local industry resulting from the Industrial Revolution.

4. A chemical is classified as a HAZMAT if it has any of the following characteristics *except*

A. exhaustibility.
B. toxicity.
C. ignitability.
D. corrosivity.
E. reactivity.

5. All of the following are methods of waste disposal *except*

A. deep-well injection.
B. landfills.
C. tectonic incineration.
D. orphan dumpsites.
E. surface impoundments.

6. Catalytic converters, scrubbers, and carbon filters are all examples of

A. pollution prevention.
B. pollution conversion.
C. pollution control.
D. pollution acceleration.
E. pollution legislation.

7. Driving hybrid cars as opposed to conventional cars is an example of

A. pollution prevention.
B. pollution conversion.
C. pollution control.
D. pollution acceleration.
E. pollution legislation.

Free Response Questions

1. Many laws have been passed to protect workers from hazardous materials, clean hazardous wastes sites, or prevent accidents involving hazardous materials. Give three examples of a hazardous waste, and name one law that would help prevent or clean up those wastes.

Multiple Choice Answers and Explanations

1. B. The RCRA states that all land disposal facilities must have a permit to operate and that toxic wastes must be pretreated to prevent leaching.

2. A. OSHA states that by law workers must have information about hazardous materials they are working with and protective equipment they should be wearing. This information is recorded on material safety data sheets.

3. D. Hooker Chemical and Plastics Company buried chemical wastes in Love Canal and then sold the land to a school board. Later, the chemical wastes leached out into the surrounding environment, causing people in the surrounding area to develop serious health problems.

4. A. HAZMATs are chemicals that are toxic, ignitable, corrosive, or reactive.

5. C. Waste disposal methods include deep-well injections, landfills, orphan dumpsites, and surface impoundments.

6. C. Filters, scrubbers, and catalytic converters are pollution controls because they are designed to help control or prevent the amount of pollution entering the environment.

7. A. Driving a hybrid car is an example of pollution prevention because it is an attempt to reduce or eliminate the production of pollution.

Solutions to Free Response Questions:

1. To be a hazardous waste, a chemical must be ignitable, corrosive to storage tanks and equipment, unstable when mixed with water, or toxic to humans if inhaled or digested. Several laws have been passed to protect people and the environment from toxic chemicals. The Resource Conservation and Recovery Act (RCRA) requires that all land disposal facilities have a permit in order to operate, that toxic wastes be pretreated before storing to prevent leaching, and that toxic substances being transported be well documented and ready for inspection at any time. The Department of Transportation Regulations are designed to help reduce the risk of spills, fires, and poisonous fumes. Because of this law, any vehicle carrying hazardous wastes must use large placards to identify the hazardous materials they are carrying and, in some cases, take designated routes on certain highways or during certain times. The Occupational Safety and Health Act (OSHA) requires that workers be given information about any hazardous materials with which they are working and the protective equipment they should be wearing to protect themselves. This information is recorded on material safety data sheets, which OSHA requires accompany the chemicals at all times.

The Atmosphere: Climate, Climate Change, and Ozone Depletion

20.1 Atmosphere and Weather

Atmosphere

The Earth's atmosphere can be divided into several layers. The two layers closest to the surface of the Earth are called the troposphere and the stratosphere. The following table summarizes the properties of these two layers.

Troposphere	Stratosphere
This layer extends from the Earth's surface to 5–10 miles high.	This layer extends from the top of the troposphere to about 40 miles high.
This layer contains water vapor and weather.	This layer contains ozone.
Temperatures decrease as you go up in this layer.	Temperatures increase as you go up in this layer.
Pollutants can be washed back to the Earth.	Pollutants remain in this layer.

Weather

Weather is the day-to-day variations in temperature, air pressure, wind, humidity, and precipitation. Weather is fueled by the sun and is affected by the rotation of the Earth on its axis. While some solar radiation is reflected by clouds or the Earth's surface, most is absorbed by the atmosphere, oceans, and land. Some of the heat generated by this absorption process is radiated back into space as infrared energy at night.

Climate is the long-term weather patterns in an area.

Meteorology is the study of weather and climate

Convection currents are vertical air currents caused by warmer air from the surface rising, expanding, cooling, and then sinking again. (Review Chapter 7.)

The **jet streams** are rivers of air high in the troposphere that flow eastward at speeds of more than 300 mph.

Fronts are boundaries of air masses with different temperatures and pressures.

20.2 Climate

While global average temperature has shown periods of cooling and warming since the 1880s, the general trend has been an increase of 0.8°C (Fig. 20–5). Records indicate that the Earth's climate has oscillated between ice ages and warm periods.

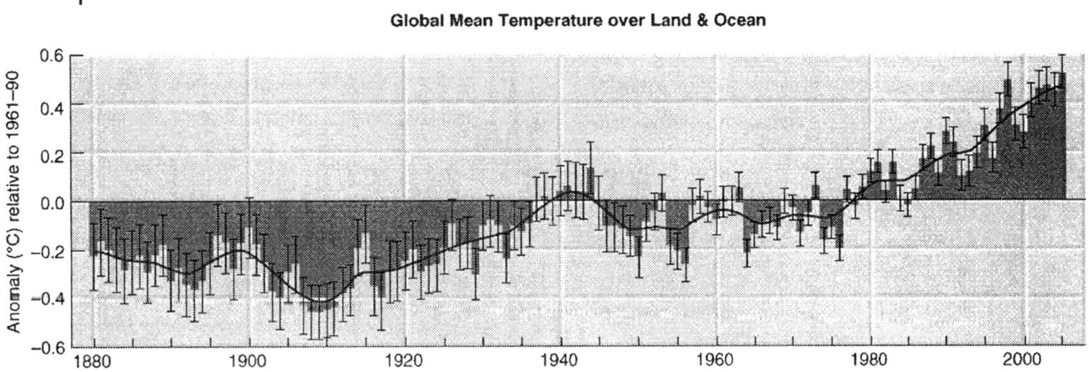

Figure 20-5 Environmental Science, 10/e
© 2008 Pearson Prentice Hall, Inc.

Two-thirds of the Earth is covered with water. Oceans play a dominant role in determining climate because they are the major source of water for the hydrologic cycle and the main source of heat entering the atmosphere (latent heat of condensation). Oceans have a high heat capacity, the ability to absorb energy when heated, and the ability to convey heat throughout the globe.

El Niño occurs when a major shift in atmospheric pressure over the central equatorial Pacific Ocean reverses the trade winds that normally blow from the east. Warm water spreads to the east, the jet streams strengthen and shift, and patterns in precipitation and evaporation are affected. Nutrient-rich cold water currents are diminished which in turn effects fishing practices.

La Niña is the opposite of El Niño. Easterly trade winds blow with great intensity and cooler water from the depths of the eastern Pacific well up to the surface and move westward, the jet streams weaken, and weather patterns are again affected. While meteorologists know *how* El Niño and La Niña happen, they do not know *why* they happen.

Thermohaline circulation is a pattern that affects ocean currents. Thermohaline refers to the effects that temperature and salinity have on the density of seawater. The thermohaline circulation pattern is like a giant conveyor belt that is circulating water masses from the surface to deep oceans and back again, depending on density and mass. This phenomenon explains why it is warmer in Europe than the latitudes would suggest.

- The Gulf Stream moves northward and is cooled by Arctic air currents.
- This cooling causes the water to be denser and thus sink.
- The deep water spreads southward to the southern tip of Africa.
- It then spreads northward, where it slows down, warms, and becomes less dense and rises again.
- This transfers heat to Europe and affects its climate.

Evidence suggests that this conveyor system could be interrupted by the addition of large quantities of fresh water in the North Atlantic (from the melting of polar ice), which would lower the density of the water. This, in turn, would prevent the sinking of the water and then block the movement of warmer water. Such an interruption to the thermohaline circulation pattern could change climate patterns with possibly disastrous effects.

20.3 Global Climate Change
As noted, the Earth has experienced periods of both warming and cooling. There are several factors that contribute to atmospheric warming and cooling.

<u>Warming Processes</u>

- **Greenhouse gases (GHGs)** contribute to atmospheric warming. Greenhouse gases include water vapor, carbon dioxide, and other gases in the atmosphere. The Earth is heated by the light it absorbs from the Sun. Some of this heat is radiated back into space, but some remains trapped in the atmosphere by gases, which act like a giant "greenhouse" covering the Earth. This greenhouse effect becomes worse as the concentration of greenhouse gases in the atmosphere increases. Figure 20–9 shows the increase in atmosphere carbon dioxide concentrations since 1960.

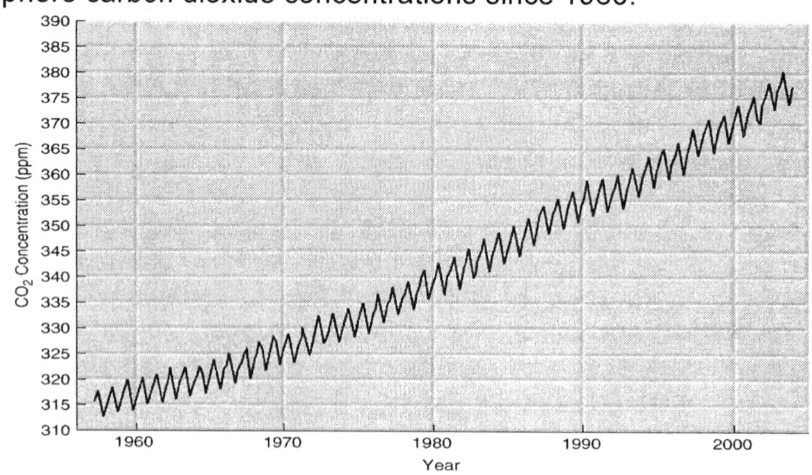

Figure 20-9 Environmental Science, 10/e
© 2008 Pearson Prentice Hall, Inc.

Cooling Processes

- **Planetary albedo** contributes to global cooling. Clouds, which cover about half of the Earth, reflect 21% of solar radiation back to space before it ever reaches the planet.

- **Volcanoes** also contribute to global cooling. When a volcano erupts, particles and aerosols enter into the atmosphere and spread around the world, causing solar radiation to be reflected back to space.

- **Aerosols** from pollution also can cause a haze in the atmosphere that reflects sunlight or increases cloud cover.

Both warming and cooling processes can change climate around the globe.

Intergovernmental Panel on Climate Change

In 1988, the United Nations Environment Program and the World Meteorological Society established the **Intergovernmental Panel on Climate Change**. This panel, which consists of more than 2,000 unpaid experts from a hundred countries, gathers and examines scientific data in an attempt to gain a better understanding of whether humans are inducing climate change.

In its 2001 Third Assessment report, the panel made it clear that they believe global climate change is happening, that this climate change is the result of an increase in greenhouse gases in the atmosphere, and that this warming trend will cause unprecedented and severe global climate changes.

In its Fourth Assessment, which was released in 2007, the panel also presented the following findings about atmospheric concentrations of carbon dioxide and other greenhouse gases.

- Carbon dioxide is increasing at a rate of 1.8 ppm/year.

- Rates of carbon dioxide vary depending on the season. (Plants are able to take in more carbon dioxide in the summer than in the winter.)

- The source of increasing carbon dioxide is the burning of fossil fuels.

- While there are natural sinks such as oceans and forests that are able to absorb carbon dioxide, these sinks are not able to keep up with the increasing amount of carbon dioxide emissions.

Other Greenhouse Gases that Lead to Global Warming

Water vapor, methane, nitrous oxide, ozone, and chlorofluorocarbons (CFCs) also contribute to the warming effect. These greenhouse gases are described in the following table.

Water Vapor	As temperatures rise, water vapor in the atmosphere increases and warming accelerates.
Methane	There are many sources of methane, including wetlands, livestock flatulence, landfills, coal mines, natural gas, rice cultivation, and manure.
Nitrous Oxide	This greenhouse gas has increased 18% from agriculture and the burning of biomass and fossil fuels. This gas accelerates warming and also contributes to depletion of the ozone in the stratosphere.
Ozone (0_3)	This potential greenhouse gas forming in the troposphere is increasing as the result of the burning of fossil fuels, forest fires, and agricultural wastes.
CFCs	CFCs contribute to global warming and to ozone destruction.

Environmental Impacts Associated with Global Warming

When the atmosphere warms, the oceans warm also. People are concerned that sea levels will rise as a result of thermal expansion of the oceans and the melting of the glaciers. Satellite data indicate that sea levels rose 15 cm during the twentieth century.

Other environmental impacts of global warming include

- an increase in intense tropical hurricanes;
- greater forest insect damage;
- an increase in insect-borne diseases like malaria;
- an increase in heat waves;
- an increase in droughts;
- a shrinking of the Arctic sea ice;
- a 20% reduction in the volume of the polar ice cap;
- an unprecedented melting of the Greenland ice sheet;

- a retreating of the glaciers;
- a change in rainfall rates;
- a migration of fish species north, which is affecting the fishing industry;
- a stronger El Niño effect;
- an increase in the flooding of coastal towns and islands.

Arctic Climate Impact Assessment

- 300 scientists met and published this assessment.
- Arctic temperatures are increasing at double the rate of the rest of the world.
- The two-mile high ice sheet in Greenland is melting and could raise sea levels by 23 feet.
- Arctic plant and animal ranges will change as a result of Arctic warming.
- Thawing ground will affect permafrost and buildings will settle and crack.

Antarctic

- Most of the world's ice is in the Antarctic.
- Many glaciers are retreating.
- Satellite measurements show that the Greenland and Antarctic ice sheets are losing as much as 36 cubic miles of ice a year.

20.4 Response to Climate Change

Mitigation Response

Mitigation refers to actions designed to reduce greenhouse gas emissions.

In 1992, at the **UNCED Earth Summit in Rio de Janeiro** (Framework Convention on Climate Change), several heads of state agreed to stabilize greenhouse gas levels in the atmosphere to 1990 levels by the year 2000. This failed.

In 1997, nations met in Kyoto, Japan, and agreed on ways to reduce greenhouse gas emissions. The agreement drafted became known as the **Kyoto Protocol**.

- The agreement is to reduce six types of greenhouse gas emissions to 5% below 1990 levels by 2012.
- Developing countries refused to sign the agreement. They argued that developed countries had caused the problems and that developing countries should be able to use fossil fuels.
- Even if all 55 countries ratify the agreement, the Kyoto Protocol will not stabilize atmospheric greenhouse gases. Worldwide emissions would have to decrease by 60% in order to have a real stabilizing effect.
- Most nations are trying to reach the goal by shifting to renewable energy sources, nuclear energy, and conservation strategies.

Adaptation Response

Adaptation means accepting that climate change is happening and taking action to lessen the impact of this change.

The poor are more vulnerable to climate change.

- Poorer countries are more susceptible to water, food, health, and disease issues than are richer countries.
- Poorer countries have fewer resources to stop the impact of climate change.
- The ecosystems the poor rely heavily upon will be disrupted by climate change.

20.5 Depletion of the Ozone Layer

The **ozone** layer acts like a shield above the Earth, protecting the Earth by absorbing high-energy radiation emitted by the Sun. Ozone is good up high but harmful at ground level. Ozone in the stratosphere protects us from UV light. Ozone in the troposphere harms us, especially children and people with respiratory problems. Ozone in the stratosphere is being depleted by chlorofluorocarbons (CFCs).

How is ozone formed and broken down?

- Ozone in the stratosphere is formed when UV light acts on oxygen, causing the oxygen (O_2) to combine with other oxygen to form ozone (O_3).

- Sometimes, oxygen molecules and free oxygen atoms reform, which means that ozone in the stratosphere is constantly forming and breaking apart.

- Chemicals in the stratosphere can upset this dynamic process and have undesirable consequences.

CFCs, a type of halogen, are one such chemical upsetting this process:

- CFCs are non-reactive, nonflammable, and nontoxic molecules that contain chlorine and fluorine atoms.

- CFCs are used in refrigerators, air conditioners, foams, electronics, and aerosol cans.

- When released into the atmosphere, CFCs mix with normal gases and eventually reach the stratosphere.

- CFCs damage the stratosphere by reacting with UV light and releasing free chlorine atoms, which attack stratospheric ozone by forming chlorine monoxide.

- In 1978, the EPA banned the use of CFCs in aerosol cans in the United States.

In 1985, scientists in Antarctica reported a thinning in the ozone layer over the South Pole. This thinning is commonly referred to as the ozone "hole." According to the EPA, the ozone hole has contributed to an increase in skin cancer and cataracts among humans.

The **Montreal Protocol**, signed in 1987,is an agreement of members of the United Nations to scale back CFC production by 50%. The aim of the protocol is to protect the ozone layer. To date, 184 countries have signed the agreement.

The United States' **Clean Air Act** also addresses the problem of stratospheric ozone depletion.

Multiple Choice Questions

1. Many times, countries have come together to discuss how to address environmental issues. An example of this is

A. the Clean Water Act.
B. CERCLA.
C. TNRCQ.
D. the Montreal Protocol.
E. RCRA.

2. Weather is most affected by the Sun and

A. global warming.
B. the tilt of the Earth's axis of rotation.
C. convection currents.
D. jet streams.
E. the melting of the Greenland ice sheet.

3. The current rise in sea level is attributed mostly to

A. the melting of the Greenland ice sheet.
B. thermal expansion of the ocean.
C. sinking of the land masses.
D. increased flooding worldwide.
E. increased precipitation in coastal areas.

4. Which of the following is true about stratospheric ozone?

A. It is dangerous to people with respiratory illness.
B. It is important technologically because it receives radio waves.
C. It is increasing due to human pollution.
D. It absorbs ultraviolet light.
E. It is being depleted due to the burning of fossil fuels.

5. Carbon dioxide levels fluctuate mostly because

A. of deforestation.
B. of ozone depletion at the poles.
C. during the winter plants lose their leaves.
D. of the rise and fall of the tides.
E. mean global temperatures.

6. Cows and other ruminants contribute to an increase in greenhouse gases by releasing

A. oxygen.
B. methane.
C. hydrogen dioxide.
D. sulfuric acid.
E. phosphorous.

7. Which of the following is likely to occur if global warming continues?

A. flooding incidences will worsen
B. diseases like malaria will decrease
C. biodiversity will increase
D. the hurricane season will shorten
E. droughts will decrease

8. Where is radon a problem?

A. in underground storage tanks
B. in coal-burning power plants
C. in soil
D. in aquatic ecosystems
E. in basements

9. What event is characterized by a major shift in atmospheric pressure over the central equatorial Pacific Ocean that reverses the trade winds and forces warmer water to spread eastward?

A. El Niño
B. La Niña
C. global warming
D. acid deposition
E. thermohaline circulation

10. Which of the following can cause the atmosphere to become cooler?

A. hurricanes
B. typhoons
C. suspended particulate matter
D. a shift in deep ocean currents
E. a change in rainfall rates

Free Response Questions

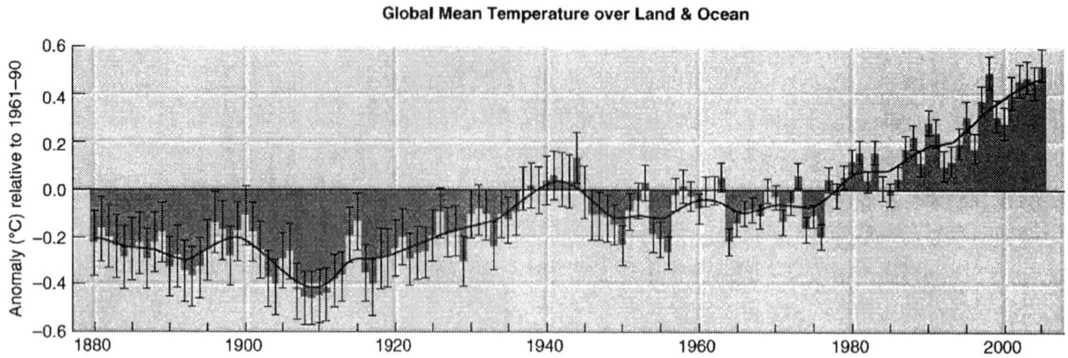

Figure 20-5 Environmental Science, 10/e
© 2008 Pearson Prentice Hall, Inc.

1. According to the graph, how much has the mean global temperature increased from 1910 to 2000?

2. List and explain three reasons for the trend in the graph.

Multiple Choice Answers and Explanations

1. D. In 1987, the United Nations agreed to scale back CFC production by 50% to reduce ozone depletion. This agreement is known as the Montreal Protocol.

2. B. Weather is fueled by the Sun and affected by the rotation of the Earth on its axis.

3. B. When the atmosphere warms, the oceans warm also. Warming of the oceans causes thermal expansion and glaciers to melt, both of which cause the level of the sea to rise. Satellite data indicate that sea levels rose 15 cm during the twentieth century.

4. D. Stratospheric ozone absorbs much of the ultraviolet light from the Sun and prevents it from entering Earth's atmosphere.

5. C. Carbon dioxide levels fluctuate in response to the life cycle of plants. During the winter, plants drop their leaves, which is where they take in carbon dioxide and release oxygen, so the amount of carbon dioxide in the atmosphere increases. The opposite is true during the summer months when plants have their leaves and are absorbing more carbon dioxide.

6. B. Animals release methane from flatulence due to anaerobic bacteria living in their gut.

7. A. As global warming continues, many scientist believe that polar ice caps and glaciers will continue to melt, which will cause oceans to expand and flooding events to increase.

8. E. Radon, a gas found naturally in bedrock, can seep into basements and cause respiratory problems in humans.

9. A. El Niño is when a major shift in atmospheric pressure over the central equatorial Pacific Ocean reverses the trade winds that normally blow from the east. Warm water spreads to the east, the jet streams strengthen and shift, and patterns in precipitation and evaporation are affected.

10. C. Particles and aerosols that enter into the atmosphere and spread around the Earth can cause solar radiation to be reflected back to space, resulting in a cooling effect.

Free Response Answers

1. Global temperature increased by 0.8°C between 1910 and 2000. In 1910, the mean global temperature anomaly was –0.4. By the year 2000, it had risen to a 0.4 anomaly, a difference of 0.8°C.

2. Many reasons have been attributed to the increase in global temperature. One reason is the burning of fossil fuels to supply energy for our vehicles and electricity. Another reason is an increase in the release of methane from anaerobic respiration in landfills, animals, and wetlands. Third, as the temperature around the globe increases, more water vaporizes, which causes a positive feedback loop and further increases temperature.

Atmospheric Pollution

21.1 Air Pollution Essentials

The atmosphere is made up of the following elements:

- nitrogen, 78%
- oxygen, 20.95%
- argon, 0.93%
- carbon dioxide, 0.04%
- water vapor, 0 to 4%
- other trace gasses

There are three things that determine the level of air pollution:

- the amount of pollution;
- the amount of space;
- what is working to remove the pollutants.

Natural air pollutants include volcanoes, fires, and dust storms.

Anthropogenic, or human-caused air pollutants, come from emissions from industrials plants, coal-burning power plants, and automobiles.

Nature can "clean up" the air if we do not overload it.

Smog is the combination of smoke and fog.

- **Industrial smog** is a grayish mixture of soot, sulfurous compounds, and water vapor found near industries (Fig. 21–3a).

- **Photochemical smog** is a brownish haze that is found in places where it is warm and sunny. This kind of smog is produced from automobile exhausts (nitrogen oxides and volatile organic carbon compounds) and sunlight (Fig. 21–3b).

Industrial smog

Figure 21-3a Environmental Science, 10/e
© 2008 Pearson Prentice Hall, Inc.

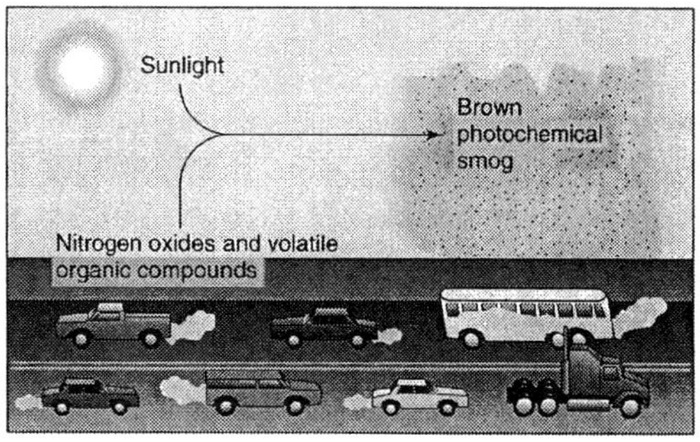

Photochemical smog

Figure 21-3b Environmental Science, 10/e
© 2008 Pearson Prentice Hall, Inc.

Temperature inversions can make smog worse.

- The sunlight hits the Earth during the day and is hottest near the ground.
- The warm air near the ground rises, carrying pollutants up into the atmosphere (Fig. 21–5a).
- A front can reverse this situation by trapping cool air below a pocket of warm air. When this happens, a temperature inversion has occurred (Fig. 21–5b).
- Temperature inversions trap pollutants close to the Earth's surface and can cause negative human health effects (e.g., headaches, nausea, eye and throat irritation).
- Temperature inversions often occur at night, and topography can make smog worse. Ex. Los Angeles

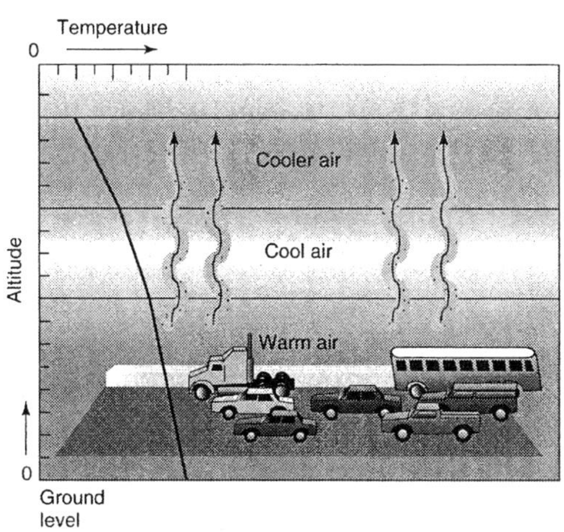

Figure 21-5a Environmental Science, 10/e
© 2008 Pearson Prentice Hall, Inc.

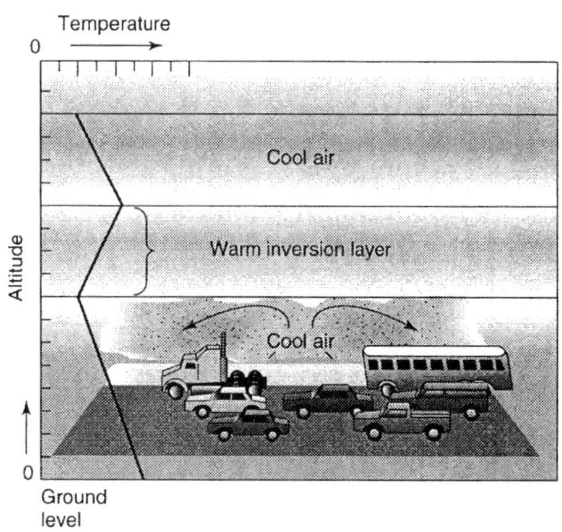

Figure 21-5b Environmental Science, 10/e
© 2008 Pearson Prentice Hall, Inc.

21.2 Major Air Pollutants and Their Sources

Primary pollutants are direct products of combustion and evaporation. The leading primary pollutants are

- particulates,
- volatile organic compounds (VOCs),
- carbon monoxide,
- nitrogen oxides,
- sulfur dioxide,
- lead,
- air toxics,
- radon.

Secondary pollutants are primary pollutants that have undergone further reactions. The leading secondary pollutants are

- ozone (reactions between VOCs and NO_x),
- peroxyacetyl nitrates (reactions between VOCs and NO_x),
- sulfuric acid,
- nitric acid.

Acid deposition is a mixture of acid rain and dry-particle fallout.

- An **acid** is any chemical that releases hydrogen.

- A **base** is any chemical that releases hydroxide ions.

- **pH** is the concentration of hydrogen ions expressed on a scale from 0 (very acidic) to 14 (very basic) with 7 being neutral. The numbers represent powers of 10, so a pH of 5 is 10 times as acidic as a pH of 6 (Fig. 21–13).

Fig. 21-13

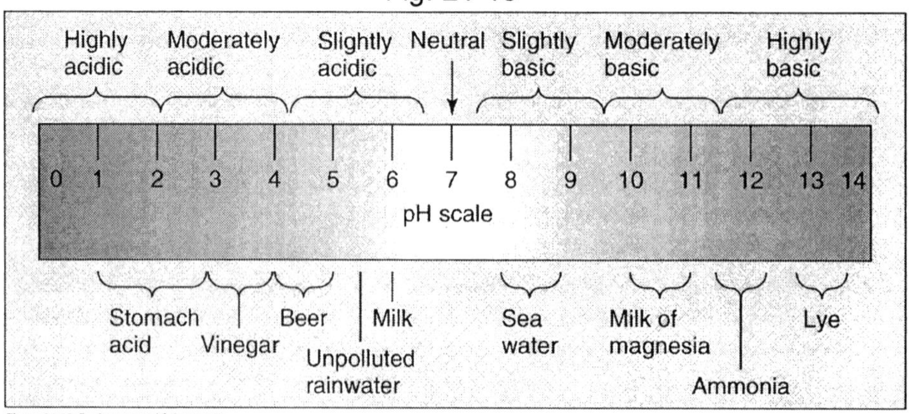

Figure 21-13 Environmental Science, 10/e
© 2008 Pearson Prentice Hall, Inc.

21.3 Impacts of Air Pollutants

Air pollutants can cause the following health problems in humans:

- COPD (chronic obstructive pulmonary disease),
- asthma,
- cancers,
- death.

Air pollutants can cause the following environmental problems:

- damage to crops;
- damage to forests;
- damage to materials like metal and stone;
- mercury accumulation in fish populations.

One way to reverse or buffer the effects of acids is to add a base. In nature, limestone is a natural buffer that protects lakes from the effects of acid rain. If acid rain continues, however, the natural buffer can be used up.

21.4 Bringing Air Pollution Under Control

The **Clean Air Act** of 1970 sets ambient standards that the most widespread air pollutants must meet.

The **Clean Air Act Amendments (CAAA)** of 1990 target specific pollutants more directly and patrol these more aggressively. These amendments also established a permit process. Polluters must apply for a permit that identifies the type of pollutants they release and the steps they are taking to lower the amount of pollution they produce.

The **National Ambient Air Quality Standards (NAAQS)** set primary standards for pollutants based on how much humans can tolerate them without ill effects. These pollutants include four primary pollutants—particulates, sulfur dioxide, carbon monoxide, and nitrogen oxides—and one secondary pollutant—ozone.

To reduce the amount of air pollution caused by particulates, the Clean Air Act banned the use of open burning and placed a "no visible emissions" requirement on industrial stacks. Many industries installed filters and other devices to meet these requirements.

New cars built after 2000 emit 75% less pollution than cars built before 1970. This reduction is due to the addition of catalytic converters to cars. Catalytic converters oxidize most volatile organic carbon compounds to carbon dioxide and water and most carbon monoxide to carbon dioxide.

CAFE or corporate average fuel economy standards, set mile per gallon (mpg) minimums on motor vehicles. For example, passenger cars must get an average of 27.5 mpg and light trucks (e.g., pickups, SUVs, minivans) must get an average of 20.7 mpg.

Multiple Choice Questions

1. Which of the following is true about acid rain?

A. It increases crop yields.
B. It destroys monuments and buildings.
C. It causes fish eggs to be thicker and enriched in calcium.
D. It leads to aquifer depletion.
E. It increases evaporation and cloud formation.

2. What is the most abundant element in the Earth's atmosphere?

A. oxygen
B. carbon dioxide
C. ozone
D. nitrogen
E. water vapor

3. What is one way to reduce atmospheric carbon dioxide?

A. decrease the use of fossil fuels
B. build taller smokestacks
C. sprinkle limestone over the pollution
D. remove catalytic converters
E. reduce dependency on mass transit

4. One gallon of gasoline produces 5 pounds of carbon dioxide. You and your friend are driving to a site 500 miles away. Your friend is driving a pickup truck, which gets 20 miles to the gallon. You are driving a new hybrid vehicle, which gets 50 miles to the gallon. By the end of the trip, how many more pounds of carbon dioxide will your friend's pickup put into the atmosphere than your hybrid vehicle?

A. 200 lbs
B. 150 lbs
C. 100 lbs
D. 75 lbs
E. 25 lbs

5. What would be the best method to decrease acid rain and acid deposition problems?

A. switch to hybrid vehicles
B. add hydroxide to reverse the effect of hydrogen ions
C. drive during daylight hours only
D. genetically alter crops to be resistant to acid rain
E. repair statues and monuments to return them to their original condition

6. Many cities have large air pollution problems because of their topography. The biggest problem in these cities is

A. acid rain.
B. smog.
C. soil denitrification.
D. habitat loss.
E. decrease in proportion of edge habitat.

7. According to the National Ambient Air Quality Standards, all of the following are classified as primary pollutants *except*

A. SO_2.
B. particulates.
C. NO_X.
D. O_3.
E. CO.

8. What characteristic of limestone makes it a natural buffer to acid deposition?

A. chromatic dispersion
B. texture
C. chemical composition
D. mass
E. eutrophication

Free Response Questions

1. Explain how a temperature inversion occurs. How would an increase in air pollution be different in the presence of a temperature inversion compared to an increase in air pollution without the presence of a temperature inversion?

Multiple Choice Answers and Explanations

1. B. Acid rain destroys monuments and statues.

2. D. Nitrogen makes up 78% of the Earth's atmosphere.

3. A. Fossil fuels are a major contributor to atmospheric carbon dioxide, so reducing the use of fossil fuels would reduce atmospheric carbon dioxide.

4. D. At 20 mpg, your friend's pickup will need 25 gallons of gasoline to travel 500 miles (i.e., 500 divided by 20 mpg equals 25 gallons), which is 15 gallons more than your hybrid vehicle will need. If one gallon of gasoline produces 5 pounds of carbon dioxide, then your friend's pickup truck will produce 75 pounds more carbon dioxide than your hybrid vehicle (15 gallons times 5 pounds per gallon equals 75 pounds per gallon).

5. A. Switching to hybrid vehicles will lower the use of fossil fuels, thereby reducing acid rain and acid deposition problems.

6. B. Smog is a problem in areas with large mountains or other topographical issues that trap pollutants near the ground.

7. D. Ozone (O_3) is a secondary pollutant.

8. C. The chemical composition of limestone is basic and therefore acts as a buffer to acid deposition.

Free Response Answers

1. Normally, sunlight is hottest near the ground. When warm air near the ground rises, it carries with it pollutants up into the atmosphere. The presence of a front reverses the situation and causes a temperature inversion: the air closest to the Earth is cooler than the air higher in the atmosphere. This temperature inversion traps pollutants close to the Earth, causing smog. This inversion often occurs at night, and an area's topography can make smog worse. Air pollution is worse during a temperature inversion because pollutants are trapped near the ground where people are breathing it rather than being spread out over large distances in the upper atmosphere. In some cases, pollution during temperature inversions can lead to accidental poisonings and death.

Economics, Public Policy, and the Environment

22.1 Economics and Public Policy

In a **command economy,** the rulers make all economic decisions.

In a **market economy,** individuals determine what is exchanged.

A **mixed economy** has elements of both a command and a market economy.

22.2 Resources and the Wealth of Nations

GNP (gross national product) is the sum of all goods and services produced *by* a country in a given timeframe. For example, GNP includes Coca Cola® bottled overseas but not Toyotas® made in the United States.

GDP (gross domestic product) is the sum of all goods and services produced *in* a country in a given timeframe. GDP is GNP minus what is being produced in other countries. For example, GDP includes Toyotas® made in the United States but not Coca Cola® bottled overseas.

22.3 Pollution and Public Policy

Policy development refers to the political/social process in response to a human or environmental problem. There are four stages to policy development:

- recognition,
- formulation,
- implementation,
- control.

22.4 Benefit-Cost Analysis

A **benefit-cost analysis** compares the cost of a proposed action (e.g., regulation, law) to its environmental benefit measured in dollars.

- If costs exceed benefits, then the proposed action is rejected.
- If benefits exceed cost, then the proposed action is accepted.
- If costs equal benefits, then the proposed actions is started /stopped.

External cost refers to something that a business does that is not included in its profit and loss (e.g., pollution).

Pollution-control costs go up as levels of a pollutant go down.

Some of the benefits associated with environmental regulation include

- improvements in human health and life expectancy;
- reductions in the corrosion and deterioration of materials;
- reductions to the damage of the environment;
- increases in aesthetic value and outdoor recreation.

22.5 Politics, the Public, and Public Policy

Environmental concerns have become a battle of special-interest groups.

The public is involved by paying higher taxes and costs on the products they consume.

Multiple Choice Questions

1. Prior to enacting an environmental law, what must policy makers do?

A. conduct a benefit-cost analysis
B. estimate profitability
C. enact trade sanctions
D. publish a findings statement
E. respond to lobbyist efforts

2. What is one way to measure the economic growth of a country?

A. fertility rate
B. birth rate
C. gross domestic product
D. infant mortality rate
E. consumer price index

3. If the cost of an environmental regulation is $4 billion, then the benefit should

A. be less than $4 billion.
B. be more than $4 billion.
C. be at least $4 billion.
D. be at most $4 billion.
E. not be considered.

4. Gross domestic product may overstate a country's level of well-being because it does not include which of the following?

A. household spending
B. depreciation of environmental capital
C. government spending
D. business investment
E. foreign trade

5. What is a possible market solution to deforestation?

A. a system of national preserves
B. laws regulating the cutting of trees
C. private purchase of forests by environmental groups
D. raising social awareness of environmental problems
E. educational efforts to promote forest friendly policies

Free Response Question

1. List two economic benefits and two economic costs to remediating the effect of soil salinization.

Multiple Choice Answers and Explanations

1. A. When doing a benefit-cost analysis, the policy maker compares the cost of the environmental law to its environmental benefit measured in dollars. If costs exceed the benefit of the law, then the law is not pursued. If the benefit of the law exceeds its cost, then the law is pursued. If the cost of the law equals its benefit, then the law is started or stopped.

2. C. The gross domestic product (GDP) is a measure of a country's economic growth. It is a country's gross national product (GNP) minus what is being produced in other countries.

3. C. If the cost of an environmental regulation is $4 billion, then the benefit should be at least $4 billion. If the benefit of the regulation does not at least equal its cost, then the regulation fails the benefit-cost analysis.

4. B. Environmental capital is never considered in a country's GDP.

5. C. By purchasing the forest, the environmental group could keep others from tearing down the trees.

Free Response Answers

1. Remediating the soil would improve the health of the soil, thereby allowing farmers to earn an income off of it. It also could increase production because the native grasses could be replanted and used for sustainable grazing. However, getting the soil back to its previous condition would require a sizable investment in time, money, and labor. Also, one would have to pay to dispose of the unusable soil.

Sustainable Communities and Lifestyles

23.1 Urban Sprawl

Urban sprawl refers to the unchecked outward growth of cities into rural or undeveloped areas. This process relocates houses, shopping centers, and workplaces from the city center to outlying areas.

Infrastructure: roads, dams, electrical systems, airports, water systems.

There are four ways to determine how "sprawled out" a city is:

- measure the density of the population.
- measure how separated the stores, homes, and workplaces are from each other.
- measure the accessibility of roads.
- measure how well the downtown area is defined and whether it is the center of activity.

<u>Urban sprawl negatively impacts the environment in the following ways:</u>

- As people move farther out of cities, they have to drive farther and thus use more oil and gas.
- More driving leads to more air pollution.
- As people move out and pave driveways and parking lots, there is increased runoff and erosion, which affects water quality.
- As people move out of the city, they move onto lands once used for agriculture.
- Wild habitats are being fragmented and destroyed as people spread out of cities.

<u>Urban sprawl negatively impacts humans in the following ways:</u>

- Each person drives farther and owns more cars.
- The risk of car accidents increases due to increased driving.
- People walk less and use mass transit less often.

- Traffic congestion is worse.
- Taxes and city services are increased.
- Obesity and high blood pressure is on the rise due to lack of physical activity.

There are some benefits of urban sprawl.

- As people spread out, residential density decreases.
- Better school and social services are available.
- Crime rates tend to drop.
- The cost of housing declines.

23.2 Urban Blight

Urban blight is the opposite of urban sprawl. With urban blight, people flood to cities looking for jobs at a rate that far exceeds the capacity of the city to absorb them. This results in urban slums. Urban blight is common in developing countries.

Urban blight leads to many negative problems, including

- a declining tax base;
- deteriorating real estate;
- declining schools and city services;
- high unemployment rate (50% or more);
- minimum wage jobs.

Urban blight does have some positive effects, including

- reduction in number of cars and an increase in walking and mass transit;
- reduction in energy consumption and pollution due to fewer cars;
- increase in recycling and composting.

23.3 Moving Toward Sustainable Communities

A sustainable city is defined as a city that has improved

- socially;
- economically;
- environmentally.

23.4 Toward the Common Good

To achieve sustainability, we must

- make decisions based on needs and values.
- use knowledge of how the natural world works.
- make economically feasible and socially desirable choices.

The following chart lists what each person can due to help reach a more sustainable world.

Change your lifestyle	Switch to a more fuel-efficient car; walk or ride a bike; recycle
Become politically involved	Vote or write your legislator
Join an environmental organization	Become aware of environmental issues and ways you can help
Join a volunteer organization	Build houses or volunteer to accomplish a goal
Make a career choice	Choose a career that solves environmental problems (e.g., lawyers, engineers, teachers, scientists)

Multiple Choice Questions

1. The environmental impacts of urban sprawl include all of the following *except*

A. loss of habitat.
B. water pollution.
C. air pollution.
D. loss of wildlife areas.
E. a declining tax base.

2. In the developing world, many people are flooding to cities looking for jobs. This phenomenon is termed

A. urban sprawl.
B. urban blight.
C. GDP.
D. suburbanization.
E. eutrophication.

3. Many inner cities have an unemployment rate of more than

A. 50%.
B. 60%.
C. 70%.
D. 80%.
E. 90%.

4. What term describes the process of people in the developed world moving out of the city and into the surrounding areas?

A. urban sprawl
B. urban blight
C. GDP
D. suburbanizing
E. eutrophication

5. Which of the following is **not** a way of helping to create a sustainable world?

A. become politically active in environmental issues
B. volunteer with environmental organizations
C. pursue a career in environmental engineering
D. bicycle to work
E. buy a large house far from your place of employment

Free Response Question

1. Each person can contribute to the common good. List three ways that you can alter the way you live to help create a more sustainable world.

Multiple Choice Answers and Explanations

1. E. A declining tax base is not an environmental impact of urban sprawl. Environmental impacts of urban sprawl include an increase in use of energy resources, air and water pollution, and loss of natural areas and agricultural land.

2. B. Urban blight is the term used to describe the phenomenon of people in the developing world flooding into cities in search of jobs at a rate that far exceeds the capacity of the city to absorb them.

3. D. Inner cities can have an unemployment rate of 50% or more.

4. A. Urban sprawl is the term that is used to refer to the phenomenon of cities growing at an unchecked rate into surrounding rural and undeveloped areas.

5. E. Buying a larger house will not help to create a more sustainable world. A larger house will use more energy to run, more resources to build, and require more land to develop.

Free Response Answers

1. There are many things that I could do to help create a more sustainable world. I could sell my car and purchase a hybrid vehicle. I can walk or ride my bike when taking shorter trips or combine many smaller trips into one. I can turn off lights when I leave a room and replace my existing lightbulbs with more energy efficient ones. I can join an environmental group and educate myself about ways that I can make a difference. I can write my legislator and share with him or her my environmental opinion and things I would like to see changed. I can recycle my waste and start a compost pile. All these things will help to make a more sustainable world for future generations.

Environmental Science

A total of 3 hours is allotted for this examination: 90 minutes for Section I, which consists of 100 multiple-choice questions, and 90 minutes for Section II, which consists of four free-response questions. Section I is printed in this examination booklet. Section II is printed in a separate booklet.

NO CALCULATORS MAY BE USED IN THIS SECTION OR IN SECTION II OF THE EXAMINATION.

SECTION I
Time- 1 hour and 30 minutes
Number of questions- 100
Percent of total grade- 60

This section of the examination contains 100 multiple-choice questions. Therefore, please be careful to fill in only the ovals that are preceded by numbers 1 through 100 on your answer sheet. NO CALCULATORS ARE ALLOWED.

General Instructions

DO NOT OPEN THIS BOOKLET UNTIL YOU ARE INSTRUCTED TO DO SO.

INDICATE ALL YOUR ANSWERS TO QUESTIONS IN SECTION I ON THE SEPARATE ANSWER SHEET. No credit will be given for anything written in this examination booklet, but you may use this booklet for notes or scratch work. After you have decided which of the suggested answers is best, COMPLETELY fill in its corresponding oval on the answer sheet. Mark only one answer for each question. If you change an answer, be sure that the previous mark is erased completely.

Example: Sample Answer

Chicago is a

(A) state
(B) city
(C) country
(D) continent
(E) village

(A) (B) (C) (D) (E)

Haphazard guessing is discouraged by a penalty of a quarter point subtracted for every wrong answer. If you leave the question blank, you will not be penalized any points. Therefore, if you cannot eliminate any choices, you should leave the answer blank, but if you can eliminate one or more choices, then it may be good to guess. Use your time wisely, and work as quickly as you can. If you come to a question that you do not know, skip it and come back to it later if you have time.

ENVIRONMENTAL SCIENCE
Section I
Time- 1 hour and 30 minutes
Part A

Directions: Each set of lettered choices below refers to the numbered questions or statements immediately following it. Select the one lettered choice that best answers each question or best fits each statement and then fill in its corresponding oval on the answer sheet. A lettered choice may be used once, more than once, or not at all in each set.

Questions 1–4 refer to sources of energy.

(A) oil
(B) lignite
(C) natural gas
(D) uranium
(E) anthracite

1. This low-grade fossil fuel is used to produce electricity.
2. This energy source is used primarily in the northeastern United States for home heating.
3. This clean burning fossil fuel is used for heat and electricity.
4. This energy source creates problems for disposal.

Questions 5–7 refer to the Earth.

(A) biosphere
(B) atmosphere
(C) hydrosphere
(D) lithosphere
(E) exosphere

5. Most chemical elements are found here.
6. This part of the Earth is occupied by living things.
7. This is a major source of hydrogen and oxygen.

Questions 8–11 refer to the carbon cycle.

(A) oceans
(B) fossil fuels
(C) atmosphere
(D) biota
(E) soils

8. This part of the carbon cycle is the largest net source of atmospheric carbon dioxide emissions.
9. This part of the carbon cycle receives carbon dioxide through the deposit of detritus.
10. This part of the carbon cycle is an anthropogenic source of carbon dioxide.
11. This part of the carbon cycle releases 60 Gt per year of carbon dioxide through plant and animal respiration.

Questions 12–14 refer to the structure of ecosystems.

(A) population
(B) community
(C) ecosystem
(D) ecotone
(E) biome

12. This is a geographic area comprised of multiple populations.
13. This is a transition area between two ecosystems.
14. Tropical rainforest, desert, and tundra are examples.

Questions 15–18 refer to air pollutants.

(A) water vapor
(B) methane
(C) nitrous oxide
(D) ozone
(E) chlorofluorocarbons

15. This banned pollutant contributes to ozone destruction.
16. This pollutant increases as temperature increases.
17. This gas is helpful in the stratosphere but harmful in the troposphere.
18. This gas is produced by landfills, sewage treatment, and agriculture.

Questions 19–21 refer to countries.

(A) United States
(B) China
(C) India
(D) Brazil
(E) Canada

19. This country has the highest per capita production of greenhouse gasses.
20. This country has the fastest growing population.
21. This country has the largest amount of total forests.

Questions 22–24 refer to population growth and carrying capacity.

(A) exponential growth
(B) S-curve
(C) J-curve
(D) carrying capacity
(E) K strategy

22. This occurs when populations level off after a period of exponential growth.
23. This is the maximum population an ecosystem can support.
24. This refers to animals that produce few offspring but many of which reach adulthood.

Directions: Each question or incomplete statement below is followed by five suggested answers or completions. Select the one letter option that is best in each case and then fill in its corresponding oval on the answer sheet.

25. What is the best measure of a country's economic growth rate?

(A) gross domestic product
(B) personal consumption expenditure
(C) consumer price index
(D) crude birth rate
(E) crude death rate

26. Which of the following processes leads to habitat destruction?

i. Conversion
ii. Fragmentation
iii. Simplification

(A) i only
(B) ii only
(C) i and ii only
(D) ii and iii only
(E) i, ii and iii

27. Which of the following is not a benefit of wind power?

(A) The land beneath the turbines can be used for farming.
(B) Wind power produces no air pollution.
(C) Wind power is free.
(D) Wind power can be used in many places around the world.
(E) Wind power is quiet and efficient.

28. The Kyoto Protocol calls for all of the following except

(A) reductions in emissions of six greenhouse gases to below 1990 levels
(B) a global collaboration for a sustainable environment
(C) allowances for carbon-credit trading
(D) money to clean up toxic waste sites
(E) exemptions for developing nations

29. During which stage of the sewage-treatment process is oxygen used to encourage growth of organic material?

(A) preliminary treatment
(B) primary treatment
(C) secondary treatment
(D) final cleansing and disinfection
(E) tertiary treatment

30. What type of pollutant can best be treated with chlorination?

(A) sediment
(B) inorganic plant nutrients
(C) organic waste
(D) pathogens
(E) chemical pollutants

31. The process of bacteria taking nitrogen gas and converting it to ammonium is called

(A) denitrification
(B) nitrogen fixation
(C) ammonification
(D) assimilation
(E) nitrification

32. Environmental problems occur when a population rapidly grows. Such environmental consequences of rapid growth include all of the following except

(A) emigration out of cities to rural farms
(B) eutrophication
(C) anthropogenic air pollution
(D) intensified land cultivation
(E) habitat loss

33. What is the net annual percentage growth rate of a population of 10,000 that experiences fifty births and thirty deaths and twenty immigrations and thirty emigrations in a year?

(A) 0.1%
(B) 1.2%
(C) 1%
(D) 12%
(E) 10%

34. A scientist finds that the LD50 of a particular rat poison kills 100 rats. What does this finding mean?

(A) The scientist started with 100 rats and all the rats died.
(B) The scientist started with 200 rats and 100 rats died.
(C) The scientist started with 1,000 rats and 100 rats died.
(D) The scientist started with 150 rats and 100 rats died.
(E) The pesticide did not work because none of the 100 rats died.

35. In the United States, most trash is

(A) put in a landfill
(B) incinerated
(C) recycled
(D) used for co-generation
(E) dumped offshore

36. Unlike active solar systems, passive solar systems

(A) are pollution free
(B) generate electricity
(C) rely on the Sun's radiant heat
(D) require photovoltaic cells
(E) require batteries to store usable electricity

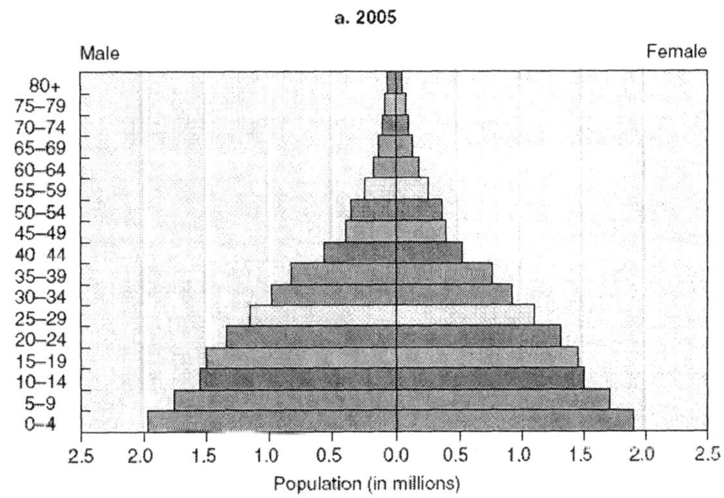

37. The country with the age distribution shown in the graph above is probably

(A) experiencing rapid growth
(B) experiencing a low death rate
(C) experiencing ZPG
(D) a developed country
(E) dealing with a large elderly society

38. Thermohaline circulation is characterized by all of the following except

(A) the Gulf Stream flowing northward where it is cooled in the Arctic
(B) the sinking of cooler waters
(C) warming near Antarctica
(D) movement northward where currents become less dense
(E) the transference of heat to Europe

39. Nonpoint sources of pollution such as fertilizers coming from Midwestern farms can lead to

(A) eutrophication
(B) soil denitrification
(C) acid rain
(D) global warming
(E) thermohaline circulation

40. Approximately what percentage of all grain crop grown in the United States is used to feed domestic animals?

(A) 10%
(B) 30%
(C) 50%
(D) 70%
(E) 90%

41. What rock contains a tar-like substance that can be heated and refined?

(A) lignite coal
(B) bituminous coal
(C) peat
(D) oil shale
(E) oil sand

42. Sand has

(A) high permeability and high porosity
(B) high permeability and low porosity
(C) low permeability and high porosity
(D) low permeability and low porosity
(E) high permeability and no porosity

43. What is not an obstacle to recycling in the United States?

(A) expense
(B) consumer apathy
(C) lack of facilities
(D) available technology
(E) political opposition

44. In which soil layer would you most likely find eluviation?

(A) A Horizon
(B) B Horizon
(C) C Horizon
(D) E Horizon
(E) O Horizon

45. The air we breathe is an example of a/an

(A) pollutant
(B) hazard
(C) common
(D) cycle
(E) antagonism

46. What species is most likely to suffer from bioaccumulation of toxic material?

(A) ant
(B) maple
(C) deer
(D) rabbit
(E) leopard

47. Which of the following gases or substances is considered a secondary pollutant in the National Ambient Air Quality Standards?

(A) nitrogen oxide
(B) sulfur dioxide
(C) carbon dioxide
(D) particulate
(E) ozone

48. Which of the following strategies does not improve a landfill?

(A) placing landfills below the water table
(B) building leachate collection systems
(C) stacking trash in a pyramid shape
(D) placing groundwater monitoring wells around the landfill
(E) covering the trash each night with soil

49. Desertification occurs for all of the following reasons except

(A) adiabatic cooling
(B) rain shadow effect
(C) drying of lakes
(D) poor irrigation practices
(E) rising sea level

50. The current human population is nearing

(A) 5 million
(B) 6.1 billion
(C) 6.6 billion
(D) 7 million
(E) 8.1 billion

51. What soil texture is best suited for agriculture?

(A) sand
(B) clay
(C) silt
(D) loam
(E) sandy clay

52. At a sewage-treatment facility, bar screens and fine screens are used to

(A) eliminate microorganisms living in the sewage
(B) settle out the grit that comes into the plant
(C) filter out the solid material that cannot be digested
(D) dechlorinate water so it can be disposed safely
(E) chlorinate water to remove bacteria

53. Ozone in the troposphere is dangerous because it

(A) blocks out the ultraviolet light
(B) is a respiratory irritant
(C) causes damage to the cells of plants
(D) is a carcinogen
(E) destroys monuments and buildings

54. A common way that radon, a known carcinogen, enters houses is through

(A) power lines
(B) basements
(C) telephone lines
(D) air conditioners
(E) heating units

Questions 55–56 refer to the following graph.

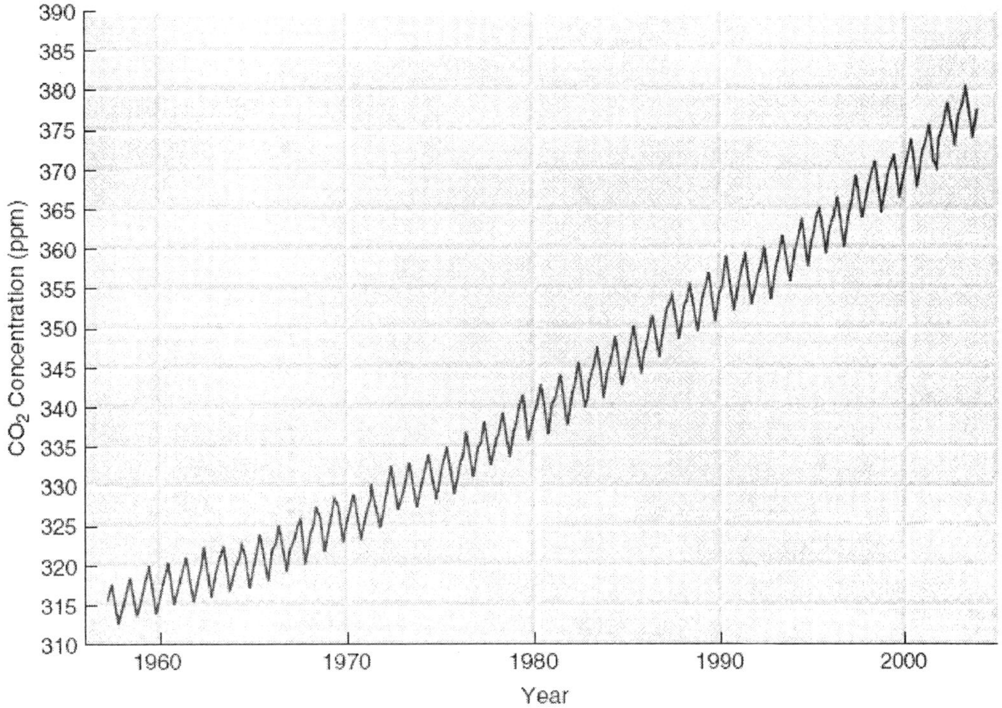

55. The graph shows that the amount of CO_2 in the atmosphere fluctuates each year. This fluctuation is due to

(A) seasonal photosynthetic activity
(B) people driving more in the summer months
(C) thermal expansion of the oceans
(D) power plants being pulled off the grid for maintenance
(E) fluctuations in volcanic activity

56. Scientists have linked the rise in the amount of CO_2 in the atmosphere from 1960 to 2000 to

(A) a rise in sea level
(B) increased El Niño events
(C) deforestation
(D) increased temperatures
(E) increased ice at the poles

57. Critics of plastic recycling cite all of the following reasons as support for their position except

(A) recovering plastic is more costly than starting from scratch
(B) plastics create no plastic leachate
(C) plastics create no dangerous biogas
(D) plastics and combustion facilities burn very hot and leave no ash
(E) plastics do not decompose

58. One problem with large-scale hydroelectric plants is

(A) controlling flooding downstream
(B) flooding upstream habitats
(C) producing air pollutants while running the plant
(D) generating a renewable water source
(E) recreational activities that the lake provides

Questions 59–60 refer to the following figure.

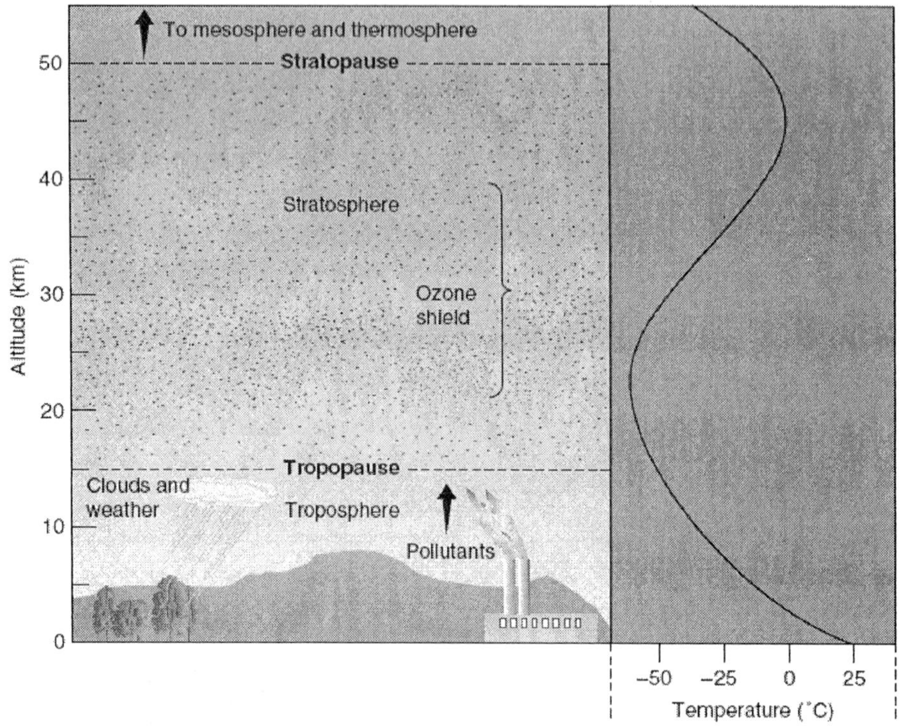

59. Which of the following statements is best supported by the data in the figure?

(A) Temperature increases as you rise in the troposphere.
(B) Temperature increases as you rise in the mesosphere.
(C) Temperature decreases as you rise in the stratosphere.
(D) Temperature decreases as you rise in troposphere.
(E) Temperature does not vary as altitude increases.

60. According to the figure, the stratosphere is about how many times thicker than the troposphere?

(A) 2 times
(B) 4 times
(C) 8 times
(D) 10 times
(E) There is not enough information to answer this question.

61. Unusually large quantities of water in the North Atlantic might cause which of the following to occur?

(A) rapid cooling
(B) rapid warming
(C) rising to the surface of colder ocean waters
(D) anthropogenic air pollution
(E) depletion of the ozone layer

62. Which of the following forms of power is a nonrenewable energy source?

(A) wind power
(B) hydropower
(C) solar power
(D) nuclear power
(E) wave power

63. Ruminants add to global warming due to their production of what greenhouse gas?

(A) NOx
(B) SO_2
(C) CO_2
(D) CH_4
(E) CO

64. If the population of a country grows at a rate of approximately 3% per year, the number of years required for the population to double is closest to

(A) 3 years
(B) 15 years
(C) 23 years
(D) 30 years
(E) 42 years

65. What federal legislation protects whales in U.S. waters?

(A) Clean Water Act
(B) Magnuson Act
(C) Whale and Dolphin Protection Act
(D) Marine Mammal Protection Act
(E) Tragedy of the Commons Act

66. One concern over global warming is

(A) extended growing seasons in parts of the world
(B) coastal towns being moved inland
(C) depletion of stratospheric ozone
(D) thermal expansion of the ocean
(E) the Sun getting closer to the Earth

67. Compared to coal-fired power plants, nuclear power plants

(A) do not require mining
(B) produce no solid wastes
(C) are not prone to accidents
(D) release carbon dioxide
(E) do not release sulfur dioxide

68. All of the following are economic costs of nuclear power except

(A) nuclear power plants do not last as long as originally thought
(B) air pollution devices
(C) production
(D) decommissioning
(E) waste disposal

69. Many people believe that the wars of the next century will be over

(A) nuclear power
(B) global warming
(C) water
(D) oil
(E) biodiversity

70. All of the following are threats to biodiversity except

(A) introduction of invasive species
(B) habitat loss
(C) human overpopulation
(D) sustainable agriculture
(E) climate change

71. An example of integrated pest management is a farmer using

(A) chlorinated hydrocarbons
(B) ladybugs
(C) DDT
(D) over-the-counter pesticides
(E) synthetic carbonates

72. The build up of pesticides over time in the tissues of animals is known as

(A) deterioration
(B) carbonate deposits
(C) biomagnification
(D) bioaccumulation
(E) biomass

73. What is an invisible, odorless, tasteless gas that binds to hemoglobin and blocks oxygen delivery to tissues?

(A) radon
(B) sulfur dioxide
(C) nitrogen dioxide
(D) carbon dioxide
(E) carbon monoxide

74. What is produced when several pollutants from automobile exhaust react with sunlight?

(A) industrial smog
(B) photochemical smog
(C) temperature inversion
(D) seasonal ambient concentration
(E) primary pollutants

75. Which of the following characteristics makes an animal more likely to become endangered?

i. small size
ii. broad distribution
iii. low reproductive rate

(A) i only
(B) ii only
(C) iii only
(D) i and ii
(E) ii and iii

76. Which of the following cycles does not have a gaseous phase?

(A) water cycle
(B) nitrogen cycle
(C) sulfur cycle
(D) phosphorous cycle
(E) carbon cycle

77. Volcanoes often have been cited as a natural cause of

(A) ozone depletion
(B) global cooling
(C) global warming
(D) thermohaline circulation
(E) increased numbers of hurricanes

78. Stratospheric ozone is referred to as the "ozone shield" because it blocks

(A) gamma rays
(B) ultraviolet light
(C) x-rays
(D) microwaves
(E) infrared radiation

79. Which of the following agricultural methods does not prevent soil erosion?

(A) contour farming
(B) shelter belts
(C) terracing
(D) overgrazing
(E) no-till agriculture

80. What phenomenon is characterized by a reversal of easterly trade winds, causing the spread of warm waters in an easterly direction and the strengthening of the jet stream?

(A) La Niña
(B) El Niño
(C) CERCLA
(D) HAZMATS
(E) OSH Act

Questions 81–82 refer to the following graph.

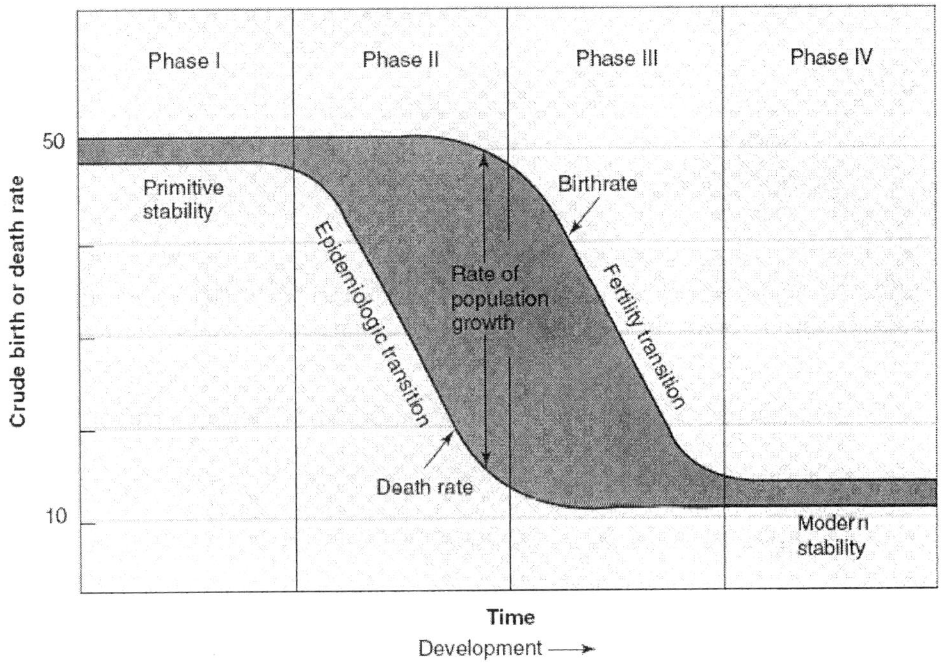

81. Zero population occurs at what point on the graph?

(A) Phase I
(B) Phase II
(C) Phase III
(D) Phase IV
(E) Phase I and IV

82. Population grows exponentially at what point on the graph?

(A) Phase I
(B) the beginning of Phase II
(C) the end of Phase II
(D) the end of Phase III
(E) Phase IV

83. Which of the following environmental problems is not caused by emissions of SO_2 and nitrogen oxide?

(A) dry fallout
(B) stratospheric ozone
(C) acid snow
(D) aluminum toxicity
(E) depletion of aquatic organisms

84. What name is used to describe the conference of nations that met to try to solve the issue of global climate change?

(A) Superfund
(B) Endangered Species Act
(C) Global Gathering on Antarctic Melting
(D) Kyoto Protocol
(E) Montreal Protocol

85. Which of the following lists the stages of coal formation in order from least heat and most moisture to most heat and least moisture?

(A) peat, lignite, bituminous, anthracite
(B) lignite, bituminous, anthracite, peat
(C) bituminous, lignite, peat, anthracite
(D) peat, bituminous, lignite, anthracite
(E) lignite, bituminous, anthracite, peat

86. What are the two most important factors used to determine climate?

(A) latitude and altitude
(B) temperature and precipitation
(C) temperature and altitude
(D) plants and animals
(E) latitude and precipitation

87. Energy being converted from one form to another is known as

(A) first law of thermodynamics
(B) second law of thermodynamics
(C) kinetic energy
(D) potential energy
(E) entropy

88. Which of the following practices is a solution to help reverse the effects of global warming?

(A) stop using chlorofluorocarbons
(B) stop overgrazing
(C) use low flow irrigation techniques
(D) switch to renewable energy resources
(E) exhaust the current supply of fossil fuels

89. All of the following are ways to decrease birth and fertility rates except

(A) reducing the literacy rate
(B) provide jobs to women
(C) provide reliable birth control
(D) provide education to women
(E) improve the status of women

90. Depletion of wetlands results in

(A) decreased sedimentation in the wetlands
(B) increased biodiversity in the wetlands
(C) increased clean water in the wetlands
(D) decreased pollution in the wetlands
(E) decreased fish stock

91. If the United States consumes 7 billion barrels of petroleum a year, approximately how many barrels per person are consumed?

(A) 10 barrels
(B) 17 barrels
(C) 23 barrels
(D) 32 barrels
(E) 47 barrels

92. What is the most likely cause of eutrophication in a body of water?

(A) run-off from fertilizers
(B) acid rain
(C) decreased salinity
(D) reproductive rate of alien species
(E) pesticides

93. Most fluctuation in annual levels of carbon dioxide in the atmosphere is caused by

(A) increased temperatures
(B) ozone depletion
(C) acid deposition
(D) thermohaline circulation
(E) seasonal photosynthetic activity

94. One way to reduce carbon dioxide emissions is to

(A) reduce dependency on mass transit
(B) outlaw catalytic converters
(C) apply limestone to affected areas
(D) decrease the use of fossil fuels
(E) increase the height of smokestacks

95. Cellular respiration contributes to levels of what gas in Earth's atmosphere?

(A) oxygen
(B) carbon dioxide
(C) methane
(D) carbon monoxide
(E) nitrogen

96. The greatest use of water in the United States is for

(A) municipal use
(B) laundry
(C) irrigation
(D) power plants
(E) industry

97. What characteristic of limestone makes it a natural buffer for acid deposition?

(A) its texture
(B) its porousness
(C) its permeability
(D) its chromatic dispersion
(E) its chemical composition

98. Where are 50% of all proven coal reserves located?

(A) Costa Rica, Thailand, and Ecuador
(B) China, Russia, and United States
(C) United States
(D) sub-Saharan Africa and New Zealand
(E) Sri Lanka and Madagascar

99. A wildlife refuge is an example of

(A) preservation
(B) conservation
(C) restoration
(D) mitigation
(E) litigation

100. Most of the Earth's atmosphere is made of

(A) carbon dioxide
(B) ozone
(C) oxygen
(D) argon
(E) nitrogen

Solutions to Multiple-Choice Practice Exam

1. B
2. A
3. E
4. D
5. D
6. B
7. C
8. A
9. E
10. B
11. D
12. B
13. D
14. E
15. E
16. A
17. D
18. B
19. A
20. C
21. D
22. B
23. D
24. E
25. A
26. E
27. C
28. D
29. C
30. C
31. B
32. A
33. A
34. B
35. A
36. C
37. A
38. C
39. A
40. D
41. E
42. B
43. D
44. D
45. C
46. E
47. E
48. A
49. A
50. C

51. D
52. C
53. B
54. B
55. A
56. D
57. E
58. B
59. D
60. A
61. A
62. D
63. D
64. C
65. D
66. D
67. E
68. B
69. C
70. D
71. B
72. D
73. E
74. B
75. C
76. D
77. B
78. B
79. D
80. B
81. E
82. C
83. B
84. D
85. A
86. B
87. A
88. D
89. A
90. D
91. C
92. A
93. E
94. D
95. B
96. C
97. E
98. B
99. A
100. E

Environmental Science

A total of 3 hours is allotted for this examination; 90 minutes for Section I, which consists of 100 multiple-choice questions, and 90 minutes for Section II, which consists of four free-response questions. Section I is printed in this examination booklet. Section II is printed in a separate booklet.

NO CALCULATORS MAY BE USED IN THIS SECTION OR IN SECTION II OF THE EXAMINATION

SECTION I
Time- 1 hour and 30 minutes
Number of questions- 100
Percent of total grade- 60

This section of the examination contains 100 multiple-choice questions. Therefore, please be careful to fill in only the ovals that are preceded by numbers 1 through 100 on your answer sheet. NO CALCULATORS ARE ALLOWED.

General Instructions

DO NOT OPEN THIS BOOKLET UNTIL YOU ARE INSTRUCTED TO DO SO.

INDICATE ALL YOUR ANSWERS TO QUESTIONS IN SECTION I ON THE SEPARATE ANSWER SHEET. No credit will be given for anything written in this examination booklet, but you may use the booklet for notes or scratch work. After you have decided which of the suggested answers is best, COMPLETELY fill in its corresponding oval on the answer sheet. Mark only one answer for each question. If you change an answer, be sure that the previous mark is erased completely.

Example:

Chicago is a

(A) State
(B) city
(C) country
(D) continent
(E) village

Sample Answer

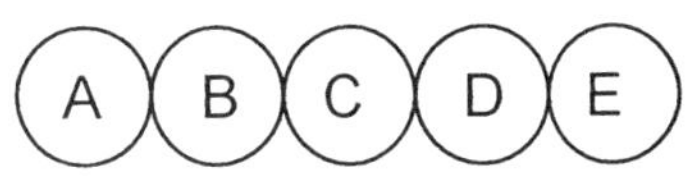

Haphazard guessing is discouraged by a penalty of a quarter point subtracted for every wrong answer. If you leave the question blank, you will not be penalized any points. Therefore, if you cannot eliminate any choices, you should leave the answer blank, but if you can eliminate one or more choices, then it may be good to guess.
Use your time wisely, and work as quickly as you can. If you come to a question that you do not know, skip it and come back to it later if you have time.

ENVIRONMENTAL SCIENCE
Section I
Time- 1 hour and 30 minutes
Part A

Directions: Each set of lettered choices below refers to the numbered questions or statements immediately following it. Select the one lettered choice that best answers each question or best fits each statement and then fill in its corresponding oval on the answer sheet. A lettered choice may be used once, more than once, or not at all in each set.

Questions 1–4 refer to renewable sources of energy.

(A) wind power
(B) solar power
(C) hydropower
(D) tidal power
(E) geothermal power

1. This renewable energy source comes from heat, deep within the Earth.
2. This renewable energy source is a result of the Coriolis effect.
3. This renewable energy source is caused by the Moon's gravitational pull.
4. This renewable energy source can be used both passively and actively.

Questions 5–7 refer to rock formations.

(A) igneous
(B) metamorphic
(C) sedimentary
(D) mineral
(E) magma

5. This is formed when rock pieces are cemented together.
6. This is formed when existing rock is under great amounts of heat and pressure.
7. This is formed when rock is melted at very high temperatures.

Questions 8–11 refer to the phosphorous cycle.

(A) eutrophication
(B) fertilizers
(C) sedimentation
(D) leaching
(E) cellular respiration

8. This causes the formation of phosphate rock.
9. This occurs when phosphorous is added to a body of water from anthropogenic sources.
10. Its overuse can lead to algal blooms.
11. This adds phosphate to the water.

Questions 12–14 refer to ecosystems.

(A) desert
(B) grassland
(C) rain forest
(D) taiga
(E) tundra

12. This biome is characterized by deep, rich, fertile soil.
13. The soil in this biome is affected by slash-and-burn agriculture.
14. Global warming will have the biggest impact on species in this biome.

Questions 15–18 refer to air pollutants.

(A) carbon monoxide
(B) volatile organic carbon compounds
(C) sulfuric acid
(D) ozone
(E) lead

15. This banned pollutant leads to nervous system damage.
16. This pollutant can lead to confusion, dizziness, and even death.
17. This is not considered a pollutant in the stratosphere.
18. This is used as a propellant in aerosol cans.

Questions 19–21 refer to countries.

(A) United States
(B) China
(C) India
(D) Brazil
(E) Japan

19. This country has a negative population growth rate.
20. This country did not ratify the Kyoto Protocol.
21. This is a low-income, developing country.

Questions 22–24 refer to the following graph.

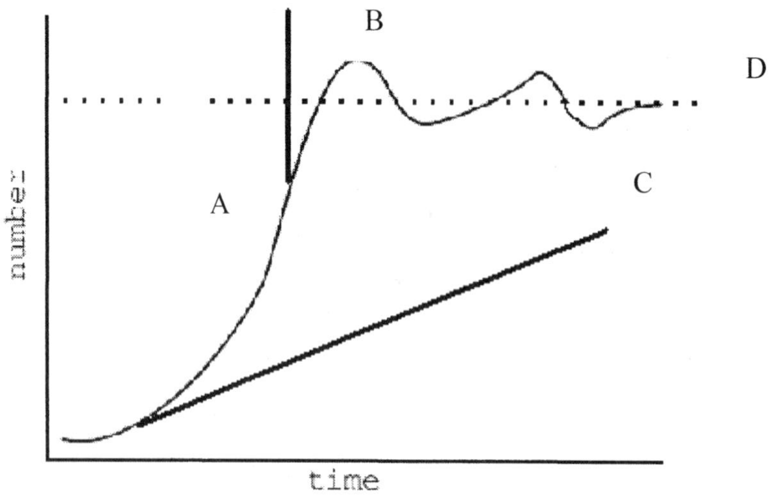

22. This letter on the graph represents exponential growth.
23. This letter on the graph represents linear growth.
24. This letter on the graph refers to the carrying capacity of the ecosystem.

Directions: Each question or incomplete statement below is followed by five suggested answers or completions. Select the one letter option that is best in each case and then fill in its corresponding oval on the answer sheet.

25. Which of the following would be the most likely to satisfy a cost-benefit analysis?

(A) using nuclear power to fuel automobiles
(B) requiring zero emissions from a coal-fired power plant
(C) disposing of hazardous waste materials on the Moon
(D) trapping methane from landfills as an energy source
(E) requiring industry to use 100% recycled materials

26. A species is less likely to become endangered if it has which of the following characteristics?

i. low biotic potential
ii. large size
iii. feeds low on the trophic pyramid

(A) i only
(B) ii only
(C) iii only
(D) i and ii
(E) ii and iii

27. Which of the following energy sources is not a feasible source of renewable energy?

(A) solar
(B) geothermal
(C) nuclear fusion
(D) hydropower
(E) wind power

28. Which of the following sources is a point source of air pollution?

(A) slash-and-burn agriculture
(B) automobile exhaust
(C) animal flatulence
(D) cigarettes
(E) volcanoes

29. Waste-to-energy generators can be used at all of the following sites except

(A) landfills
(B) sewage treatment plants
(C) cattle farms
(D) sugar cane crops
(E) limestone quarries

30. If wastewater is not treated properly, all of the following can occur except

(A) people can get sick
(B) erosion can be increased
(C) pathogens can get into a water supply
(D) eutrophication can occur
(E) point-source pollution can increase

31. What is the usual outcome of the introduction of an invasive species into an ecosystem?

(A) The ecosystem thrives.
(B) The ecosystem collapses.
(C) Biodiversity is lost.
(D) Native species weed out the invasive species.
(E) The native and invasive species develop a symbiotic relationship.

32. All of the following are social determinates that increase birth and fertility rates except

(A) higher educational attainment for women
(B) lack of reliable birth control
(C) low status of women
(D) religious beliefs
(E) living in a rural area

33. What is the net annual percentage growth rate of a population of 10,000 that experiences seventy births and thirty deaths and ten immigrations and forty emigrations in a year?

(A) 0.1%
(B) 1.2%
(C) 1%
(D) 12%
(E) 10%

34. A scientist finds that the LD50 of a particular rat poison kills 500 rats. What does this finding mean?

(A) The scientist started with 500 rats and all rats died.
(B) The scientist started with 500 rats and 50 rats died.
(C) The scientist started with 50 rats and 500 rats died.
(D) The scientist started with 150 rats and 100 rats died.
(E) The pesticide did not work because none of the 500 rats died.

35. What is an example of a prevention approach to pollution?

(A) scrubbers
(B) catalytic converters
(C) Clean Air Act
(D) hybrid cars
(E) recycling

36. Why are dams beginning to be taken down?

(A) They produce no air pollutants.
(B) They flood areas downstream.
(C) Habitats are destroyed in areas near dams.
(D) The water that can be used for drinking water is contaminated.
(E) The Clean Water Act has banned dams.

a. 1985

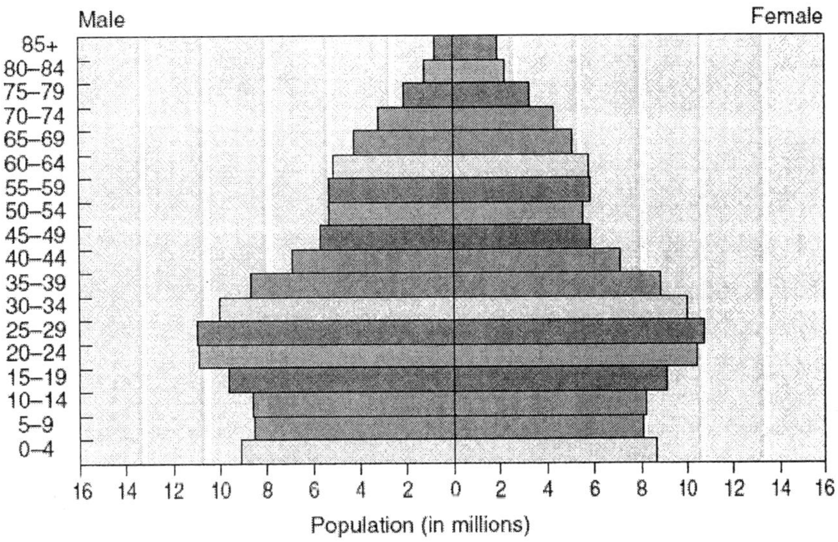

37. The country with the age distribution shown in the graph above is probably

(A) experiencing rapid growth
(B) experiencing a low death rate
(C) experiencing ZPG
(D) a developed country
(E) dealing with a large elderly society

38. Why does the amount of carbon dioxide in the atmosphere fluctuate with the seasons?

(A) Many trees drop their leaves in the winter and therefore take in less carbon dioxide.
(B) More people drive in the winter.
(C) More people farm in the summer.
(D) The troposphere increases in thickness during the winter and provides more protection from carbon dioxide.
(E) More people begin exercising when it gets warmer and exhale more carbon dioxide.

39. Soils found in tropical and subtropical rain forests that are not very fertile are called

(A) mollisols
(B) oxisols
(C) alfisols
(D) aridisols
(E) parasols

40. What soil type is best suited for agriculture?

(A) sandy loam
(B) loam
(C) clay
(D) sand
(E) silt

41. What rock, when heated to 1,100 degrees Fahrenheit, can be used to extract a type of crude oil that can be refined into gasoline?

(A) lignite coal
(B) bituminous coal
(C) peat
(D) oil shale
(E) oil sand

42. Clay has

(A) high permeability and high porosity
(B) high permeability and low porosity
(C) low permeability and high porosity
(D) low permeability and low porosity
(E) high permeability and no porosity

43. Waste-to-energy generators are

(A) at landfills and sewage treatment plants
(B) a way to recycle plastics
(C) cause fuel costs go up
(D) used in developing countries as a supplemental energy source
(E) used to power hybrid cars

44. Which of the following characteristics is not associated with tropical rain forests?

(A) biodiversity
(B) fertile soil
(C) little seasonal variation in temperature
(D) abundant precipitation
(E) abundant shade

45. The ocean is an example of a

(A) pollutant
(B) hazard
(C) common
(D) cycle
(E) antagonism

46. DDT would have the largest impact on which of the following organisms?

(A) ant
(B) Tree
(C) deer
(D) aardvark
(E) leopard

47. Which of the following statements is true of the troposphere?

(A) It contains 1,000 times more ozone than the stratosphere.
(B) It contains 1,000 times less ozone than the stratosphere.
(C) It has substances that absorb 99% of the UV light that hits the Earth.
(D) It is made up of chemicals that lead to global warming.
(E) It gets warmer as you increase in altitude.

48. What do clay liners in landfills help to reduce?

(A) air pollution
(B) ozone depletion
(C) global warming
(D) habitat destruction
(E) water pollution

49. Which of the following does not contribute to desertification?

(A) irrigation practices
(B) habitat destruction
(C) pesticides
(D) farming practices
(E) global warming

50. The current human population of the United States is nearing

(A) 5 million
(B) 300 million
(C) 6.6 billion
(D) 7 million
(E) 8.1 billion

51. The Aral Sea is considered one of the world's worst environmental disasters. This disaster was due to a large

(A) water diversion project in an area of the former Soviet Union
(B) dam used to transport ships upstream on a river in China
(C) nuclear power plant disaster
(D) brownfield area that was cleaned and restored
(E) dead zone found in thc Aral Sea basin

52. Solid wastes that can not be digested are captured in which phase of sewage treatment?

(A) primary treatment
(B) secondary treatment
(C) tertiary treatment
(D) both primary and secondary treatments
(E) all three phases capture solid wastes

53. Which of the following statements is true of ozone?

i. Ozone is the same as atmospheric oxygen.
ii. Ozone is a pollutant in the troposphere.
iii. Ozone filters ultraviolet radiation in the stratosphere.

(A) i only
(B) ii only
(C) iii only
(D) i and ii
(E) ii and iii

54. Particulate air pollution has been linked to all of the following except

(A) global warming
(B) global cooling
(C) acid deposition
(D) increased incidence of respiratory disease
(E) property damage

Questions 55–56 refer to the following graph.

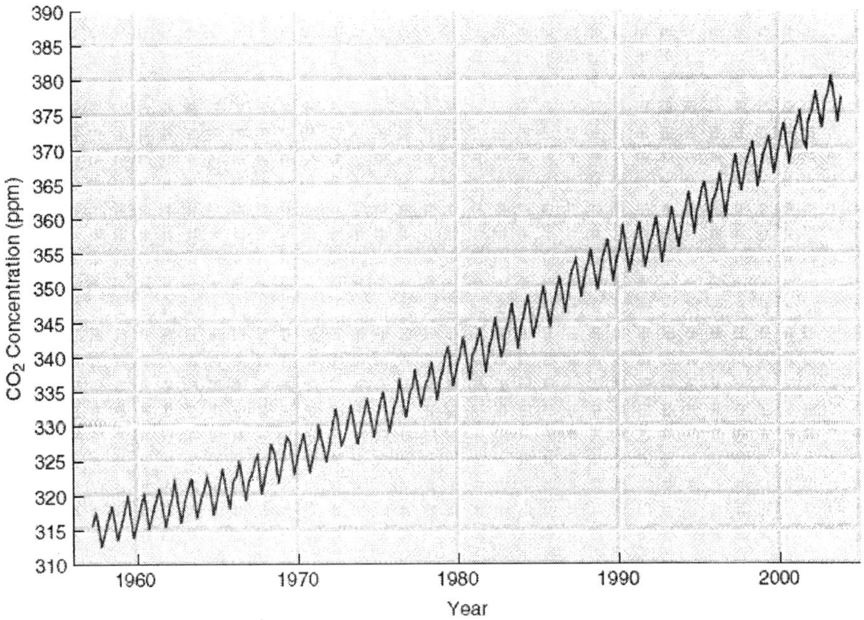

55. The graph shows that the amount of CO_2 in the atmosphere fluctuates each year. This fluctuation is due to

(A) seasonal photosynthetic activity
(B) people driving more in the summer months
(C) thermal expansion of the oceans
(D) power plants being pulled off the grid for maintenance
(E) fluctuations in volcanic activity

56. Scientists have shown that global temperature during the past 100 years has

(A) decreased by 0.5 degrees Celsius
(B) decreased by 1 degrees Celsius
(C) increased by 0.5 degrees Celsius
(D) increased by 1 degree Celsius
(E) not changed

57. Which of the following statements is true of tire recycling?

i. It reduces the risk of mosquito-born diseases.
ii. It provides material for road surfaces.
iii. It increases the risk of landfill fires.
iv. It can be used for fuel.

(A) i only
(B) i, ii, and iii
(C) i, ii, and iv
(D) ii and iii
(E) ii, iii, and iv

58. What energy source supplies most of the world's energy?

(A) solar
(B) hydroelectric
(C) wave
(D) oil
(E) nuclear

Questions 59-60 refer to the following figure.

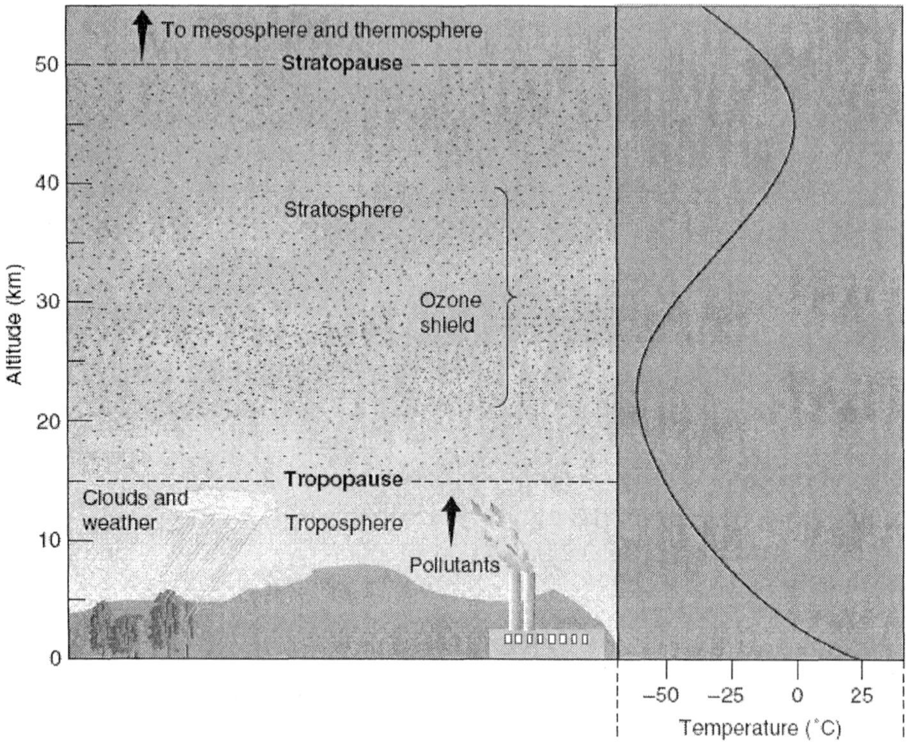

59. Which of the following statements is best supported by the data in the figure?

(A) Temperature increases as you rise in the troposphere.
(B) Temperature increases as you rise in the mesosphere.
(C) Temperature decreases as you rise in the stratosphere.
(D) Temperature decreases as you rise in troposphere.
(E) Temperature does not vary as altitude increases.

60. According to the figure, where are you most likely to encounter –10 degrees temperatures?

(A) 10 km
(B) 20 km
(C) 30 km
(D) 35 km
(E) 45 km

61. High temperatures are most likely to increase the rate of which phase of the water cycle?

(A) precipitation
(B) evaporation
(C) transpiration
(D) surface runoff
(E) condensation

62. The negative aspects of nuclear power include all of the following except

(A) human health risks
(B) environmental risks due to an accident
(C) air pollution
(D) the high cost of building and maintaining the plant
(E) environmental risks with mining uranium

63. Landfills add to global warming due to their production of what greenhouse gas?

(A) NO_x
(B) SO_2
(C) CO_2
(D) CH_4
(E) CO

64. If the population of a country grows at a rate of approximately 7% per year, the number of years required for the population to double is closest to

(A) 10 years
(B) 15 years
(C) 23 years
(D) 30 years
(E) 42 years

65. Eagles are federally protected in the United States because of the

(A) Clean Water Act
(B) Endangered Species Act
(C) Large Predatory Bird Act
(D) Kyoto Protocol
(E) Tragedy of the Commons Act

66. Global warming is linked to all of the following except

(A) nitrous oxides
(B) carbon monoxide
(C) carbon dioxide
(D) water vapor
(E) chlorofluorocarbons

67. Compared to nuclear power plants, coal-fired power plants

(A) are prone to accidents
(B) do not cause air pollution
(C) have a history of controversy due to solid-waste disposal
(D) produce most of the United States' electricity
(E) provide material for weapons manufacturing

68. What is an example of a market approach to controlling air pollution?

(A) government regulation requiring the use of scrubbers
(B) a tax on polluters
(C) government take over of a polluting factory
(D) tradeable pollution permits
(E) government regulation to reduce automobile emissions

69. Seasonal variability in carbon monoxide emissions can be attributed to what natural process?

(A) volcanoes
(B) water cycle
(C) phosphorous cycle
(D) photosynthesis
(E) thermohaline circulation

70. Why is tropical deforestation more detrimental to the environment than temperate deforestation?

(A) Soil in the tropics is extremely fertile and capable of sustaining large-scale agriculture.
(B) There is more biodiversity in the tropics than anywhere else on Earth.
(C) Temperate forest hardwoods are less valuable than tropical hardwoods.
(D) Less people rely on tropical forests than on temperate forests.
(E) Tropical forests produce less carbon dioxide when they burn than do temperate forests.

71. An example of integrated pest management is a farmer using

(A) chlorinated hydrocarbons
(B) pheromones
(C) DDT
(D) over-the-counter pesticides
(E) synthetic carbonates

72. Which of the following is not an argument for a DDT ban?

(A) The use of DDT has unknown side effects on humans.
(B) DDT causes the eggs of birds to become fragile.
(C) DDT can bioaccumulate in the tissues of animals.
(D) DDT is a form of environmental racism.
(E) DDT harms beneficial organisms as well as dangerous pests.

73. What is an invisible, odorless, tasteless gas that has carcinogenic effects?

(A) radon
(B) sulfur dioxide
(C) nitrogen dioxide
(D) carbon dioxide
(E) carbon monoxide

74. Which of the following practices is the best method for decreasing acid rain and acid deposition problems?

(A) switch to hybrid vehicles
(B) add hydroxide to reverse the effect of hydrogen ions
(C) drive during daylight hours only
(D) genetically alter crops to be resistant to acid rain
(E) repair statues and monuments to return them to their original condition

75. Which of the following characteristics makes an animal more likely to become endangered?

i. produce many offspring
ii. commercially valuable
iii. large size

(A) i only
(B) ii only
(C) iii only
(D) i and ii
(E) ii and iii

76. Metamorphic rocks are formed when

(A) sediment is carried and deposited downstream
(B) a volcano erupts and the magma hardens
(C) sedimentary or igneous rocks are put under heat and/or pressure
(D) the bodies or organisms die, decay, and are compressed
(E) when weathering wears down existing rock

77. Airplanes' vapor trains have been cited as a human cause of

(A) ozone depletion
(B) global cooling
(C) global warming
(D) thermohaline circulation
(E) increased numbers of hurricanes

78. Ultraviolet light is absorbed by what gas in which layer of the atmosphere?

(A) ozone in the troposphere
(B) carbon dioxide in the stratosphere
(C) ozone in the stratosphere
(D) nitrogen in the troposphere
(E) carbon dioxide in the troposphere

79. Which of the following choices does not accelerate the rate of soil erosion?

(A) temperature
(B) wind
(C) precipitation
(D) farming
(E) vegetation

80. What phenomenon is characterized by intense easterly trade winds that cause cooler waters from the depths of the eastern Pacific to well up to the surface and move westward?

(A) La Niña
(B) El Niño
(C) CERCLA
(D) HAZMATS
(E) OSH Act

Questions 81–82 refer to the following graph.

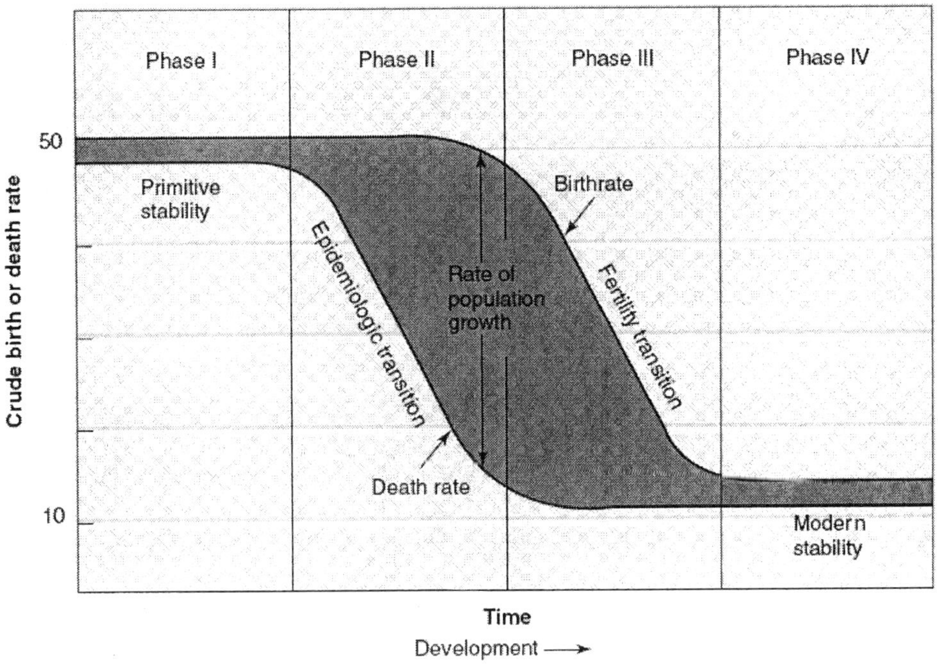

81. Explosive population growth occurs at what point on the graph?

    (A) Phase I
    (B) the beginning of Phase II
    (C) the end of Phase II
    (D) the end of Phase III
    (E) Phase IV

82. At what point on the graph is population stable?

    (A) Phase I
    (B) the beginning of Phase II
    (C) the end of Phase II
    (D) the end of Phase III
    (E) the beginning of Phase III

83. What is one way to reduce atmospheric carbon dioxide?

    (A) reduce dependency on mass transit
    (B) reduce use of hydroelectric power
    (C) ban hybrid vehicles
    (D) discourage people from using pollution credits
    (E) reduce use of fossil fuels

84. What name is used to describe a conference of nations that met to try to solve the issue of global climate change?

(A) Superfund
(B) Endangered Species Act
(C) Global Gathering on Antarctic Melting
(D) Kyoto Protocol
(E) Montreal Protocol

85. What stage of coal formation has the least amount of moisture content?

(A) peat
(B) lignite
(C) bituminous
(D) subbituminous
(E) anthracite

86. What are the two most important factors used to determine climate?

(A) latitude and altitude
(B) temperature and precipitation
(C) temperature and altitude
(D) plants and animals
(E) latitude and precipitation

87. Energy cannot be created or destroyed is known as

(A) first law of thermodynamics
(B) second law of thermodynamics
(C) kinetic energy
(D) potential energy
(E) entropy

88. Many scientists believe that if global warming continues

(A) droughts will decrease
(B) thermal expansion of the oceans will increase
(C) polar ice caps will grow
(D) sea levels will drop
(E) glaciers will advance

89. Developing countries have

(A) large GDPs
(B) large meat consumption
(C) high infant mortality rates
(D) many laws concerning environmental health
(E) large per capita incomes

90. Introduction of invasive species into an ecosystem results in

(A) species diversity
(B) species decline
(C) increased fertility
(D) long-term biodiversity
(E) a fragmented edge effect

91. What is an example of cogeneration?

(A) using solar energy
(B) building a dam for hydroelectric energy
(C) creating energy from a landfill
(D) a nuclear power plant
(E) natural gas

92. Increasing the amount of fertilizers that are applied to crops can cause

(A) eutrophication
(B) soil denitrification
(C) soil salinization
(D) resistance to plant diseases
(E) decreased pest attack

93. High-level radioactive wastes are stored at

(A) Yucca mountain
(B) Chernobyl
(C) landfills
(D) nuclear reactor sites
(E) Three-Mile Island

94. Using a ladybug to eat aphids is an example of

(A) pesticide use
(B) bioengineered crops
(C) the Green Revolution
(D) integrated pest management
(E) plant biotechnology

95. Where does most of the natural gas that Americans use come from?

(A) America
(B) Canada
(C) Russia
(D) Antarctica
(E) Persian Gulf region

96. The greatest use of water in the world is for

(A) municipal use
(B) laundry
(C) irrigation
(D) power plants
(E) industry

97. What phenomenon is characterized by a flooding of people to cities in search of jobs?

(A) urban sprawl
(B) urban blight
(C) GDP
(D) suburbanization
(E) mitigation

98. According the National Ambient Air Quality Standards, all of the following gases or substances are primary pollutants except

(A) particulates
(B) ozone
(C) carbon monoxide
(D) sulfur dioxide
(E) nitrogen oxides

99. When farmers want to increase the ability of their soils to hold water, they add

(A) clay
(B) humus
(C) loam
(D) organic fertilizers
(E) inorganic fertilizers

100. Approximately what percentage of the water on Earth is salt water?

(A) 97%
(B) 93%
(C) 87%
(D) 75%
(E) 70%

Solutions to Multiple-Choice Practice Exam

1. E
2. A
3. D
4. B
5. C
6. B
7. A
8. C
9. A
10. B
11. D
12. B
13. C
14. E
15. E
16. A
17. D
18. B
19. E
20. A
21. C
22. A
23. C
24. D
25. D
26. D
27. C
28. E
29. E
30. B
31. C
32. A
33. A
34. A
35. D
36. C
37. D
38. A
39. B
40. B
41. D
42. C
43. A
44. B
45. C
46. E
47. B
48. E
49. C
50. C

51. A
52. A
52. E
54. A
55. A
56. C
57. C
58. D
59. D
60. E
61. B
62. C
63. D
64. A
65. B
66. E
67. D
68. D
69. D
70. B
71. B
72. D
73. A
74. A
75. E
76. C
77. B
78. C
79. E
80. A
81. C
82. A
83. E
84. D
85. E
86. B
87. A
88. B
89. C
90. B
91. C
92. A
93. D
94. D
95. A
96. C
97. B
98. B
99. D
100. A